# Technology and Management

## By the same author

The Techniques of Production Management (1971)
Production and Operations Management (1979) (*4th edn. 1989*)
Management and Production (1972) *(2nd edn. 1980)*
Women in the Factory *(with A. B. Hill and C. C. Ridgeway)* (1970)
Mass Production Management (1972)
Principles of Modern Management *(with B. Lowes)* (1972) *(rev. edn. 1977)*
Work Organization (1975)
Concepts of Operations Management (1977)
Operations Management—A Policy Framework (1980)
Essentials of Production and Operations Management (1980) *(3rd edn. 1990)*
Management and Production—Readings (1981)
How to Manage (1983) *(US edn. 1985)*
Four children's books comprising the 'Read and Explain' series (1982)
International Handbook of Production and Operations Management (ed.) (1989)

# Technology and Management

Edited by

*Ray Wild*

MSc, PhD, MSc, DSc, CEng, WhF, FIMechE, FIProdE, CBIM, FRSA

*Henley—The Management College*

CASSELL

**Cassell Educational Limited**
Artillery House
Artillery Row
London SW1P 1RT



First published 1990

**British Library Cataloguing in Publication Data**
Technology and management.
1. Industries. Technological innovation. Management
I. Wild, Ray
658.5′7

ISBN 0-304-31732-2

Typeset by Area Graphics Ltd, Herts.
Printed and bound in Great Britain
by Mackays of Chatham

# Contents

# *Introduction*

The effective deployment, exploitation and management of new technology is recognised as being of critical importance in all aspects of business and industry. The potential impact of new technologies on products, processes and services is a matter of strategic significance. The importance of this will surely continue to increase.

Technology is influencing jobs and organisations. The nature of organisations is changing. Organisational structures are changing. Training needs, career paths, communications, roles and relationships within organisations are affected by technological change. This impact is unlikely to diminish. Furthermore, technology is increasingly having a major impact on the nature of management. The use of new technology is a major challenge to managers. The ability to use the 'technologies of management' effectively will be a key to business and industrial success. The importance of this must increase.

This book addresses three areas:

1. The management of technology.
2. Technology and the organisation.
3. The technology of management.

It seeks to provide an understanding of some of the issues and activities which are, or will be, of relevance in most types of business. It is an introductory book intended to be of value to those acting as technologists and managers, as well as those who can claim both roles and allegiances.

Although in a few countries there has traditionally been a significant flow of technologists into management and through to senior management roles, this is not evident in the UK, the Commonwealth countries or North America. Nor have these latter countries benefited from educational practices that have brought together the study of business and technology. In the UK, for example, recognition of the need for engineering degree courses to cover business topics has only recently been achieved.

Few business degree courses cover much technology, and hardly any MBA or executive programmes offer much in this area. So, with few exceptions, the understanding of both technology and business, and professional competence in these areas, is rarely a characteristic of one person. It follows, in these situations, that those who are required to operate in roles that span both areas must acquire the competence in some other way. Whether or not this is a satisfactory situation is not a matter that will be addressed here. The task of this book is to provide for the repair of the situation—that is, to provide a means for individuals to bridge the gap.

For many this 'gap' is a chasm. For them, this is a sensitive area. Managers without science or technology backgrounds are often apprehensive, and wary of the need to learn more about technology in order to cope in their changing roles. Avoidance behaviour is often evident. For technologists entering managerial roles, there is often evidence of a blind faith in 'the technology', and an inadequate understanding of the human factors, organisational, personnel and managerial implications. Here the risk is that of over-assertion, prejudgement and precipitative action. These two things are mutually aggravating—the one exacerbates the other. Clearly there is a need for a balance of understanding and appreciation of technology and management.

We deal here with both 'directions' of the technology/management relationship. This is not a manager's guide to technology, nor an introduction to management for the technologist. We shall deal with the common ground, or frontier, of this relationship. This, we believe, is a 'no-man's land' at the present time. Our intention is to facilitate the colonisation of this large, and important, tract from both sides, in order to encourage the development of individuals who are comfortable and confident in the whole territory on which future business success may be built. We might describe this territory in the manner of the matrix below.

| | | Management | |
|---|---|---|---|
| | | C. Managing technology | D. Using technology to manage |
| Technology | A. Specific | (i) | (ii) |
| | B. Generic | (iii) | (iv) |

Row A refers to specific technologies, i.e. those associated only or largely with a particular industry. Biotechnology, nuclear and space technology can be seen (for our purposes) as specific. Row B refers to those technologies that have relevance, use and impact that are not industry specific—the generic technologies. Information technology is the best example. These are also often seen as the 'enabling technologies'. Column C is to do with the management of technology, while column D refers to the deployment of technology in the management process. Of the four 'cells' that are the intercepts of the two sets of dimensions, we have chosen, here, to deal primarily with (iii) and (iv). We shall not be concerned with specific technologies *per se*. In an introductory book we cannot deal adequately with industry-specific technology management issues (i). In fact we shall deal with specific technologies only briefly when discussing investments in technology, and briefly also in respect to their uses by management (ii).

It will be clear that we are adopting a broad definition of technology. In fact we shall use this term to cover the hardware, software and systems utilised by an organisation as well as that incorporated in its products and services. We deal, of course, with the new technologies. This, with our focus on generic technology, implies a concentration on the enabling and information-based technologies, including computing, communications technologies, advanced manufacturing technologies and systems, automation, office technologies, etc. Unfortunately there is no simple and singular definition, but the common components of any definitions that we might use are computing, information and communications. This is our territory.

This territory is of significance to those in public and private sector organisations, whether engaged in manufacturing or in the provision of services, since such sectoral distinctions are largely irrelevant in technological development and implementation. While, often, we shall deal with the technology of products, we shall not focus only on manufacturing organisations, since one company's products may be the means by which another provides a service. We shall refer to 'companies' and to 'industries', but this should be taken as a convenient terminology rather than any suggestion of particular or limited relevance. New technologies are as relevant to the hospital administrator as to the factory manager, while technologists have similar roles and responsibilities in public transport utilities and in electronics companies.

The book is arranged in three parts, as envisaged above:

Part 1. The Management of Technology.
Part 2. The Impact of Technology on Management and Organisations.
Part 3. The Technologies of Management.

There is no sequential dependence between parts, or between chapters in each part. The only claim made is that together these 11 chapters, each prepared by an expert in the area, cover the principal aspects that are of importance to managers and technologists in most businesses.

# Biographical Notes on Contributors

**Keith Alsop** BSc, CPhys, FInstP, FIOD

A physicist by education, Keith Alsop worked first in industry and then for the government until 1978, when he took early retirement in order to pursue his interest in research management training and became a Senior Visiting Research Fellow with the Management Programme within Brunel University. Since then, his main concern has been in running a two-week course on the management of R & D three or four times a year and a number of in-house courses stemming from this.

He has been a regular chairman of selection panels for scientists for government laboratories and has served on the committee of the R & D Society.

**Peter C. Bell** MA, MBA, PhD

Peter Bell is Professor of Management Science and Information Systems at the University of Western Ontario in London, Ontario, Canada.

He is co-author of *Statistics for Business with MINITAB: Text and Cases* (1987) and *Statistics for Business with Lotus 1–2–3: Text and Cases* (1989), published by the Scientific Press, and author of over thirty articles in professional journals. He is an Associate Editor for *INFOR*, and a member of the editorial board of the *European Journal of Operational Research*.

Professor Bell is a past President of the Canadian Operational Research Society (1985–1986), and has been elected North American Vice President of the International Federation of Operational Research Societies for 1989–91.

**Adrian Campbell** BA, PhD

Adrian Campbell is a Research Fellow at the Business School, University of Aston (Birmingham), where he is currently researching the diffusion and implementation of production management systems in the engineering components industry. In 1987 he was visiting fellow at the International Institute of Management, WZB, West Berlin, where he has also carried out contract research in industrial relations for EEC-funded projects co-ordinated by the IAS Institute. His consultancy activities have included work for ICI

Films and Mars Confectionery. He is co-author of *Microelectronic Product Applications in Great Britain and West Germany* (Gower, 1989).

**Frank W. Dewhurst** BSc, PGCE, MA(Econ)
Frank Dewhurst is currently a lecturer in Quantitative Techniques and Operational Research at the Manchester School of Management, UMIST (University of Manchester Institute of Science and Technology). His research activities have centred around computational approaches for facilitating management decision-making with particular emphasis on simulation studies in production and operational management. He is now investigating the application of AI techniques, along with more traditional approaches, for such problems.

**Gordon Edge**
Professor Gordon Edge is Chief Executive Officer of Scientific Generics, the major Cambridge-based technology and business consulting and investment group. He was founder of PA Technology in 1969 and a member of the Board of PA from 1973 to 1986. He is a member of the Ericsson Science Council in Stockholm and a member of the Board of Michael Peters plc.

He is also a visiting professor at Brunel University.

**Keith Hodkinson**
Keith Hodkinson was a lecturer in Law at Manchester University before joining the Manchester office of Marks & Clerk, a partnership of Chartered Patent and Trade Mark agents, as legal consultant. He is the author of *Protecting and Exploiting New Technology and Design* (Spon, 1987) and *Employee Inventions* (Longman, 1986), and co-author of *Industrial Espionage* (Longman, 1986). His other research includes project work for the DES and DTI in technology exploitation and in legal barriers to public acceptance of new technology. He is a regular speaker at the Licensing Executives Society and other conferences on licensing law and practice.

**Rangasami L. Kashyap** DIISc, ME, PhD
R. L. Kashyap is Professor of Electrical Engineering and Associate Director of the NSF-supported Engineering Research Center on Intelligent Manufacturing at Purdue University, Indiana. He is author of *Stochastic Dynamic Models from Empirical Data*, published by Academic Press, and of more than 200 papers and book chapters in the areas of system identification, pattern recognition, image processing and intelligent manufacturing systems. Professor Kashyap was elected a Fellow of the IEEE in 1979 for his contributions in pattern recognition and stochastic automata.

**Soundar R. T. Kumara** BEng, MTech, PhD
Soundar Kumara is an assistant professor in the Department of Industrial and Management Systems Engineering at Pennsylvania State University. His research interests include artificial intelligence, database systems and pattern

recognition. Professor Kumara is a senior member of the Institute of Industrial Engineers and a member of Sigma Xi.

**Alan W. Pearson** BSc
A. W. Pearson has been Director of the R & D Research Unit at the Manchester Business School since its inception in 1967. Prior to that he worked first in industry and then in education. He is Senior Lecturer in Decision Analysis at the School and an active participant on a wide range of management and educational committees, a member of the editorial board of *IEEE Transactions on Engineering Management* and Editor of *R & D Management*. He is co-author of *Mathematics for Economists* (revised edition Macmillan, 1983) and co-editor of *Transfer Processes in Technical Change* (Sijthoff, 1978) and *Managing Interdisciplinary Research* (Wiley, 1984). A member of the boards of the International Association for the study of Interdisciplinary Research and the College of R & D of the Institute of Management Science, he is also an Adjunct Program Associate at the Center for Creative Leadership, North Carolina.

**David Preedy** BSc, MSc
Since 1983 David Preedy has been a Director of the management consultancy firm Metapraxis, responsible for the development of the Executive Information System RESOLVE, and the VISION software for information media control in a boardroom environment. He is particularly concerned with developing technology tools for directors that enhance their role and are simple to use.

**Allen L. Soyster**
Allen L. Soyster is Professor and Head of the Department of Industrial and Management Systems Engineering at Pennsylvania State University. His research interests are in the areas of mathematical programming, production planning, energy modelling, artificial intelligence and robotics. Professor Soyster is an active member of the Institute of Industrial Engineers and has served as chairman for several committees in his chapter. He is a member of the Operational Research Society of America and the Institute of Management Science. He is also a visiting lecturer for the Society of Industrial and Applied Mathematics. He is currently the Senior Editor of *IEE Transactions*.

**Brian C. Twiss** MA, MSc, CEng
Having spent some years in industry and education, Brian Twiss is currently an independent consultant in technology management and strategic forecasting. He is also Industrial Reader in Technology Management at the International Management Centres—Europe. His main field of interest is the profitable use of technology through effective R & D management, strategic forecasting and long-range planning. His current research interest is the strategic management of technical change.

Brian Twiss has written articles for many journals and is author or co-author of eight books, including *Managing Technological Innovation* (3rd edition

Pitman, 1986) and *Managing Technology for Competitive Advantage* (Pitman, 1989).

He has been a member of the Management of Engineering Committee of the Institute of Electrical Engineers, is Associate Editor (Europe) for the *Journal of Product Innovation Management* and serves on the editorial committees of the journals *R & D Management, Research Policy* and *Technology Analysis and Strategic Management*. From 1986 to 1989 he was research coordinator for the Joint Economic and Social Research Committee/Science and Engineering Research Committee programme on the successful management of technical change.

**Malcolm Warner** MA, PhD(Cantab)

Malcolm Warner has taught and researched at universities and business schools on both sides of the Atlantic. Until 1987 he was Professor of Management Studies and Research Director of the Joint Graduate Programme, Henley Management College and Brunel University. He is currently a visiting Fellow at Wolfson College, Cambridge, and a visiting faculty member of the Management Studies Group, Department of Engineering, Cambridge University. The author, co-author and editor of over fifteen books and more than one hundred papers on management and manpower, has also worked as a consultant for several business and governmental organisations, including most recently the Forecasting and Assessment of Science and Technology (FAST) of the European Commission. His most recent book (with Dr Adrian Campbell and Dr Arndt Sorge) is *Microelectronic Product Applications in Great Britain and West Germany* (Gower, 1989).

**Ray Wild**

Until recently Professor Ray Wild was Head of the Department of Manufacturing and Engineering Systems at Brunel University in West London, England. He was also pro-Vice Chancellor of the university. From April 1990 he is Principal of the Henley Management College—the oldest-established business school in Europe. He holds Master's degrees in Management and Engineering, a PhD in Management, and a DSc. He is a Fellow of two engineering institutions as well as a Companion of the British Institute of Management and Fellow of the Royal Society of Arts. His background is in engineering, technology and management studies; he has spent many years in universities and business schools as well as having worked in industry both in design and research capacities. He is the author of fourteen books and editor of four others (excluding this one), and author of over 120 papers, all on aspects of management, manufacture and technology.

# PART 1

# *The Management of Technology*

Here we shall deal with the management of technology and technology-related aspects within the organisation. We shall consider those topics that have a general relevance in industry and in business. We deal both directly and indirectly with new technology. In discussing business strategies, investment in technology and aspects of technological change, we will be considering technology *per se*. In addition, we will consider the management of those activities that give rise to technological outcomes for use within the organisation, and/or exploitation by that organisation. Thus, in our discussion of the management of scientists and technologists and the management of research and development, our approach is indirect, but of no less importance. Similarly, we shall consider the management of intellectual properties i.e. inventions, patents, designs, etc., which form the basis of most future scientific and technological developments.

# 1.1 *Business Strategies for New Technologies*

*Brian C. Twiss*

## The Corporate Role of Technology

In the final analysis the contribution of technology can only be assessed in relation to the extent that it furthers the prosperity of the organisation it serves. This chapter will explore a number of concepts and approaches that can aid the decision-maker to achieve this objective. It must be recognised that there are no simple answers, for there is no general agreement about the most effective procedures. Nevertheless, the concepts described in this chapter provide essential aids to the decision-maker in all cases. In this introduction the corporate role of technology will be explored from three different viewpoints.

While the emphasis in the chapter will be upon private sector industrial organisations producing goods, the content will also apply to public sector organisations and to service industries. Throughout, the term 'products' will be used to mean the offerings of the organisation—whether physical products or services.

### *Strategy-driven technology*

The need for strategic thinking has been widely appreciated in recent years and most large companies now have some form of strategic planning. The rationale for this is that it is essential to carry out a systematic analysis of all the forces shaping the business environment in order to establish where opportunities and threats might arise. When these are compared with the actual or potential capabilities of the firm it is possible to formulate a purposeful and coherent set of policies to guide the company into the future. This is essentially an intellectual exercise that must be conducted against the background of the corporate culture, which reflects the history of the company and the aspirations and attitudes of its top management.

Success is derived from the market-place where the company is usually only one of a number of players. Thus the strategy of any organisation must be measured against the current and future actions of competitors, which will

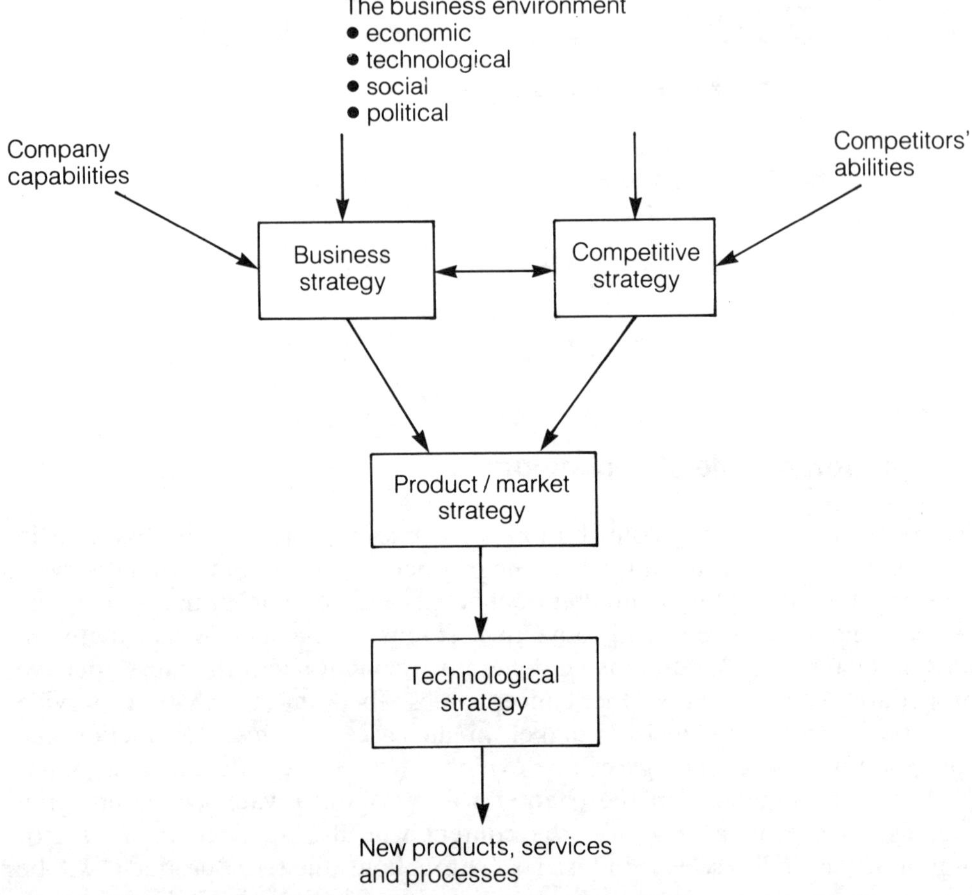


**Figure 1** Strategy-driven technology—the top-down approach.

derive from their own corporate strategies. This leads to the concept of competitive strategy.

While the strategy aims to provide guidance for the major decisions it can only be fulfilled by the operational success of the products (or services) it offers in aggregate. Sales from these must generate a higher return than their costs of provision. However, since most companies offer a range of products at different stages in their life cycles and in a variety of markets, the products and markets are to some degree mutually dependent. They must be considered as a portfolio of activities. This leads to the concept of a product/market strategy to give shape to the portfolio and integrate it with the business strategy.

The ability to earn a satisfactory financial return depends upon the inherent characteristics of the products. They must appeal to customers through their performance, price or a combination of both. Furthermore, profit is highly dependent upon the unit cost of provision. It is in achieving the objective of profit that technology makes its contribution, embodied in the product or the processes by which it is provided. For technologically based companies this is largely the responsibility of R & D. However, the work of R & D consists of a portfolio of projects that are to some extent mutually dependent, analogous but not identical to the market portfolio. Thus there is a need for a technical strategy to integrate the technology with the business and marketing strategies.

The reasoning described above leads to a logical hierarchy of strategies (Figure 1) derived from a careful analysis of the factors determining success. It is an intellectual process designed to give coherence to decision-making throughout the organisation and underlies much that is written in the business literature. Technology is implicitly regarded as playing a supportive or subservient role and is confined within boundaries determined from above. Communication is essentially from the top downwards.

## *The technology-driven approach*

Examination of the technological innovation literature suggests a very different model of how technology can contribute to corporate success. This stresses the importance of creativity and the actions of individual technical champions, who push their ideas through to success, often in the face of strong opposition elsewhere in the company.

The strategic model is criticised on the grounds that it is highly deterministic and demands a degree of technological certainty that cannot be ascribed to any proposals that are truly innovative. It is argued that the formal procedures of strategy formulation and planning found in the larger companies are likely to lead to institutionalisation and an inability to exploit the uncertain world of new technology. This results in a climate conductive to incremental improvement but hostile to genuine innovation. The achievement of small entrepreneurial ventures rather than the established industrial leaders in developing new technologies, for example in microelectronics and biotechnology, is cited. The logic of this argument is that the role of technology is so important but its path so uncertain that the corporate aim must be to exploit those ideas emerging from the technical departments and shape them to the corporate needs. This flow of ideas must be from the bottom upwards.

This line of reasoning does not necessarily reject the concept of a technical strategy. However, its aims are different. The purpose is to devote the resources to develop technical trajectories and to provide a stream of products, some radical, some incremental, to optimise the commercial output from the technological investment. This is deployed to build the technical base in a way that will lead to the development of innovative products. This does not imply that financial and market considerations should be ignored, merely that progress should be technology-driven rather than marketing-driven.

## *The Japanese approach to new technology*

Many Japanese companies adopt a radically different approach to the development of new technologies. They invest considerable resources to the setting of long-range technological objectives for up to 25 years ahead. Hitachi, for example, has a think tank of 80 full-time members for this purpose, which leads them to devote some of their research to biotechnology, on the view that this may well replace microelectronics in many applications in the long term. Having established these objectives, they make resources available to R & D with a minimum of financial evaluation. The philosophy is that if the technologist is given clear long-term objectives following a rigorous analysis he or she should then be left relatively free to take those decisions for which only he or she is competent.

### Discussion

These three different approaches have been briefly described in order to draw attention to the complexities of the strategic application of technology. They are not mutually exclusive and suggest that management should aim to achieve a blend of:

- systematic analysis leading to an understanding of the underlying forces determining business success, expressed in terms of corporate and competitive strategies;
- recognition of the realities of the innovation process in respect of the exploitation of new technologies and their influence on strategic potential;
- a long-term view based on detailed technology forecasting because of the lengthy period between the recognition of the strategic potential of a new technology and its commercial exploitation.

## The Technological Strategy

The purpose of a technological strategy is to provide a stream of products through time to further the organisation's business and competitive strategies. In practice it determines the allocation of resources between a variety of technical activities. Since the funds available are limited, choices must be made between these applications, each of which may appear attractive in its own right. The considerations to be taken into account when making these choices will be discussed later. First we must examine what these alternatives are.

### Corporate applications of technology

#### Maintenance of the current product line

Of overriding importance are likely to be the maintenance of a competitive edge and the extension of the product lives of the current products; these are likely to absorb a high proportion of the technical effort. Many of the improvements will be incremental, based upon the existing technological knowledge base. Some of them are likely to involve the application of new technologies—such as new materials and electronic management systems in the motor car. In this situation it is necessary to build expertise in a technology of which the company may have little or no past experience. Today most companies find themselves exposed to a larger portfolio of technical competences than has been the case in the past. This knowledge must be identified, evaluated, captured and transferred into the firm. This cannot be left to chance and the development of the technological knowledge base must receive systematic attention and be managed carefully. This is a strategic consideration, even for companies whose products are not particularly innovative.

## Development of new products—non-radical

All products eventually reach the end of their commercial life. This necessitates the development of a replacement. It will be based on a new design that must incorporate the latest technology, but not necessarily new technology. It should not be assumed, however, that the new product ought to be a direct replacement of the old. Developments in the market place may suggest that it should be aimed at a different market segment. In recent years most markets have experienced a growth in segmentation, with many of the most profitable products arising from high value added niche segments. To a large extent it is technology that adds the value. Any changes require a close integration of technical, market and business strategies, which should be carried out before initiation of the development of a replacement product.

## Development of new products—radical

From time to time the emergence of a new technology or a substantial performance improvement in an existing technology will cause a major disruption. A good example is the substitution of many mechanical or electromechanical products by microelectronics. This provides an opportunity for companies possessing the new technology but poses a threat to those in the technology being replaced. History indicates that the firms possessing expertise in the new technology are likely to succeed at the expense of those who have dominated the market in the past.

It is essential to identify the potential of these technologies and, more importantly, to assess *when* they will be commercially competitive. Technology forecasting can be an invaluable tool in assessing this time dimension. Even when the current market leaders have recognised the threat they may be unable to avert it without a major strategic reorientation. For a variety of reasons they may be unable to build up the necessary expertise themselves in spite of a major investment in the new technology. Thus the emergence of a new technology may necessitate the examination of radical strategic alternatives, such as the take-over of another company or a merger to acquire the new technology.

### Manufacturing process development

The processes used in the manufacturing industries represent a major application of technology, indeed the industrial revolution was characterised by the development of new methods of production rather than by new physical products. Today investment in new manufacturing technology is required to:

- enable the development of new processes to produce physical products in marketable quantities, for example new materials or biotechnological systems, previously only available on a laboratory scale;
- reduce manufacturing cost through the design or purchase of new equipment or modifications to existing plant, often incorporating new technology;
- improve the overall efficiency of the total system in relation to the needs of the business—quality, market and design responsiveness, or flexibility.

Traditionally investment in new product development and in manufacturing has not been closely integrated. It has been possible to consider R & D strategy and manufacturing strategy as largely independent. This is no longer the case, largely due to the many applications of information technology that make it essential to consider the two aspects within an integrated whole, namely a technological strategy. Research has shown that many manufacturing investments and their implementation (e.g. AMT) have not been closely integrated with the business objectives.

### Technical problem-solving

From time to time operational problems in product design or manufacturing processes will necessitate a diversion of technical resources from longer-term strategic activities. This is inevitable but must be closely controlled; the solution is a managerial rather than a strategic consideration.

## *Market and technological portfolios*

A frequently used method for illustrating a company's competitive position is the representation of its products (or strategic business units) on a 2 × 2 or 3 × 3 matrix of which the axes are 'market share' and 'growth potential' (Figure 2*a*). A similar analysis can be mde for the company's portfolio of technical activities, where the axes are 'pace of technical advance' and 'technological position' (Figure 2*b*).

There are important differences between these two portfolios. The market portfolio is a representation of the current position and the expected future growth of existing and future products within the present state of knowledge. Many of the factors determining growth in the market will not be technologically based, for example demographic, economic or fashion trends.

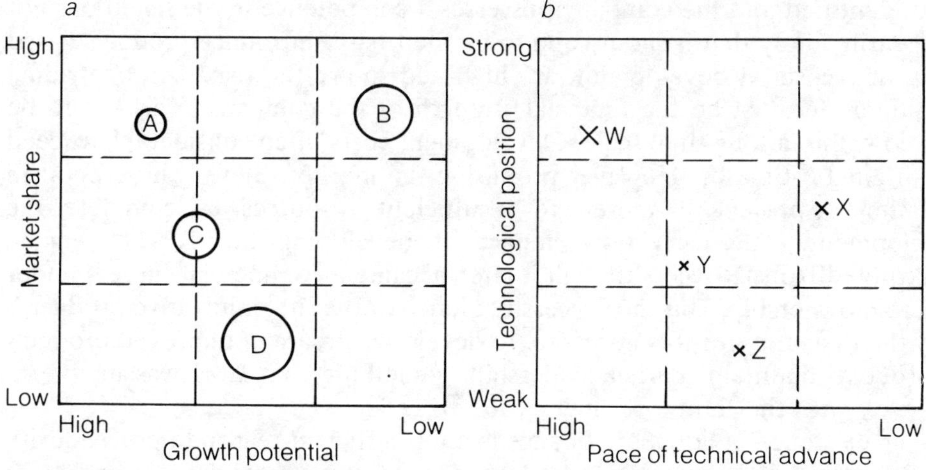


**Figure 2** Market and technological portfolio analysis. *a*, Product portfolio (the size of the circles represents either turnover or profit). *b*, Technical portfolio.

The technology portfolio relates the strength of the company's ability to the potential of a technology and is likely to reflect a longer time scale. The fact that it has a leading position in an emerging technology does not necessarily mean that it will lead to the development of products with a significant market potential within the company's strategic horizon. This dilemma could be resolved if it were possible to assess accurately the impact of a new technology. Although technology forecasting can assist, history often shows that the successful applications of a new technology occur in areas that are not foreseen. A classic example is the case of computers, where the world market was considered at one time to be numbered in single figures rather than the mass market that eventuated.

It can be seen that these considerations can lead to serious differences of opinion within a firm. Although new technology cannot be supported just because it is there and the company possesses a competence in the field, it can be shortsighted to reject it solely on the grounds that no immediate application can be identified. The purpose of a technological strategy must be to serve the two aims of furthering the existing business strategy and maintaining a coherent and purposeful technological activity within a strategic framework. The use of these matrices assists management in visualising the portfolios and provides a valuable structure for a debate to resolve the differences of view.

## *Timing of introduction*

One of the most difficult aspects in developing a technological strategy relates to timing. Methods for the assessment of the rate of advance of a new technology will be explored later. The fact that a technological potential has been identified and the company possesses a competence in the field does not necessarily imply that it should aim to be the first to introduce products based on it. The costs of development are high and so are the risks. An overriding consideration must be the financial strength of the company. This has to be viewed within a long time scale. Development costs often considerably exceed the original estimates. However promising the market, there is little point in initiating a project if there are insufficient resources to complete the development. The cash flow impact of developing the RB-111 engine bankrupted Rolls-Royce although it later became a commercial success under different ownership. But the successful launch of the first innovative product is only the first step, for it is necessary to develop a stream of improved products in order to maintain market leadership, something that EMI was unable to achieve with the body scanner.

The existence of adequate finance is not a sufficient reason to proceed with development. It is possible to adopt one of two strategies—offensive or defensive. An offensive strategy is aimed at being first to the market with the new technology and to maintain this technical leadership. This must be based upon the ability to achieve the technical objectives. It is also likely to be a high-risk strategy, although the benefits from success can be considerable. A more cautious approach, a defensive strategy, is to allow competitors to take the initial risk and to respond strongly if their development is a success, perhaps by developing a second-generation product. This involves a greater marketing risk and might be justified for a company with a strong market base. In both cases the new technology cannot be ignored, as an effective defensive strategy depends upon the ability to make a rapid technical response. In general, an offensive strategy is likely to be appropriate when the firm's distinctive competence lies in technology, where the technical risks can be assumed to be less, and a defensive strategy when the distinctive competence lies in marketing.

## *Incremental development versus radical innovation*

An existing product represents a considerable investment in design, development, manufacturing and marketing. It is highly desirable that this investment earns a return for as long as possible. This can be achieved by marketing a stream of products of improving performance and by modifications to widen appeal to a greater range of market segments. This policy cannot be maintained indefinitely. A time will come when it becomes essential to develop an entirely new design or to adopt new technologies for a radically different product.

The timing of these decisions is crucial for success. Many companies rely upon incremental improvements to their old designs long after they have begun to lose market share or to invest in new products based on an old technology in the face of competition from a new technology. This is often called the 'sailing ship syndrome', because great performance improvements in sailing ships were produced as a result of heavy investment by the industry in the face of competition from steam ships. There is much evidence that this is a common response, ranging from the investment in canals after the arrival of the railways to valve technology after the first transistors appeared and, more recently, in electromechanical systems, for example National Cash Register, in the face of competition from microelectronics. On the other hand, there are highly innovative companies that have launched a succession of entirely new products without exploiting the full commercial potential of any of them through incremental improvement.

Thus an important element of any technological strategy must be the striking of a balance between incremental product improvement and the timing of the introduction of entirely new products, often based upon new technology.

## *Discussion*

A number of considerations that must be taken into account when formulating a technological strategy have been noted. The main features are:

- The conflicting nature of the alternative demands placed upon limited technical resources. Thus choices have to be made. Because of the requirements to maintain competitive strength in the short term much of the technical budget is pre-empted, meaning that the justification for investment in new technology for the long term must be soundly based and well argued.
- The importance of timing in determining the scale and nature of the resources devoted to new technology and the launching of products based on it.
- The possibility of conflict between the needs of maintaining a long-term technical competence and the more immediate demands of the current business strategy.
- The need to develop a coherent technological strategy based upon a systematic analysis of the needs of the business, its market and the potential of new technology.

The remainder of this chapter will be devoted to an examination of concepts and techniques that aid the decision-maker in the resolution of these dilemmas.

## Industry development and technology

Many industries and companies within them are closely associated with one core technology. Although this is a simplification, since many companies are

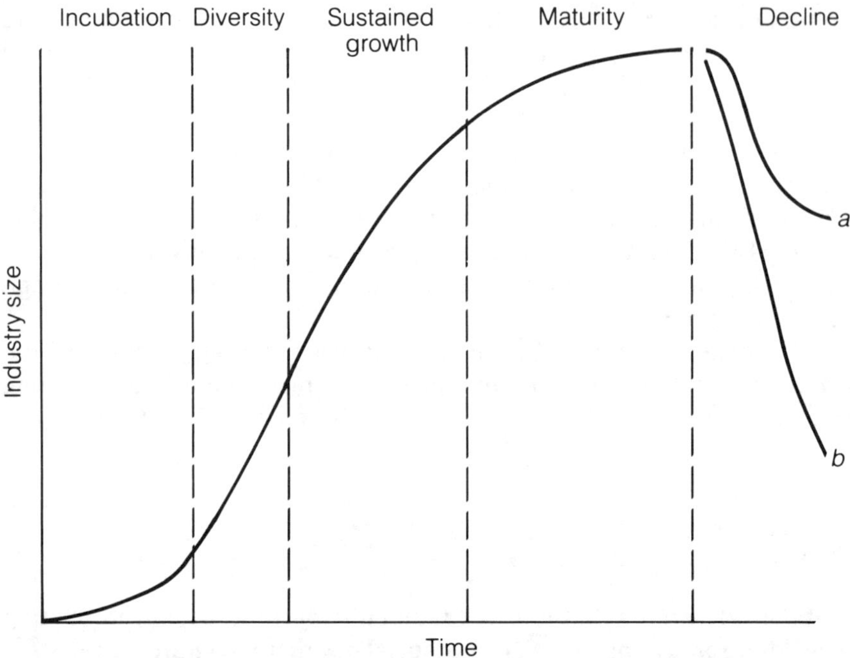


**Figure 3** The industry life cycle. *a*, Decline to a lower but sustainable level due to increased product longevity in a saturated market. *b*, Technical decline brought about by the emergence of a new technology.

technologically diversified and most products incorporate more than one technology, it provides a useful starting point for studying the development of an industry and the contribution of technology to growth. It is an important element in the formulation of a technological strategy. The typical pattern for the growth of an industry follows a curve of the shape shown in Figure 3. This can be conveniently broken down into five stages, each marked by a different emphasis in the contribution that technology can make to corporate prosperity.

## Stage 1—incubation

In the early stages of a new technology its impact is ill-defined. There is a recognition that it may have a commercial application but it is not clear where this might be. There may, however, be a limited demand in a specialised market, for example defence, where its properties enable it to gain a foothold in spite of high cost and technical problems. Usually major investments have to be made for relatively minor improvements in performance before it reaches a stage where it can make a significant profit contribution. This may cover several decades, as exemplified by carbon fibres and robotics, the potential of which was recognised in the early 1960s.

### *Stage 2—diversity*

For those technologies where the potential is sufficient to attract continued research funds a stage is reached when they become attractive in a range of applications, still often specialist. The performance of the products based upon the technology improves rapidly over a relatively short period of time and the sales volume of the companies in the industry begins to grow at a high rate. This period is characterised by a diversity of design approaches and a large number of firms, many small, engaged in the developments. Successive products with significantly enhanced performance, but with short product lives, appear in quick succession. They are often based on radically different technical approaches. Corporate success is highly dependent upon technical achievement in new product development, where the leading company gains its competitive advantage from superior technical performance. Profitability is likely to be associated with early introduction rather than low development cost. Towards the end of this stage a dominant technical configuration begins to emerge.

### *Stage 3—market growth*

With the emergence of a dominant design the pace of technical performance growth is likely to slow. In contrast, this is the period when the market shows the most marked expansion as the technology becomes widely accepted. It becomes less easy to establish a position of market dominance based solely on the basic application of the technology and there are likely to be a number of firms offering broadly the same product. The customer has a wider choice and seeks the product that best suits his or her individual needs. Although entirely new products still appear, the main emphasis shifts to the design of applications to meet the needs of specific market segments. Thus the technological emphasis is on incremental improvements, punctuated by occasional more radical innovations, closely co-ordinated with a careful analysis of market needs. Product cost becomes more important as a determinant of the buying decision. The most profitable companies are often those that apply their technology to add value.

### *Stage 4—maturity*

As the technology matures and the market approaches saturation it becomes increasingly difficult to differentiate products in terms of their technical performance. The products of competitive companies are essentially the same, and the differences are largely cosmetic. They take on many of the characteristics of a commodity in which the purchasing decision is largely determined by price. In this highly competitive market profit margins tend to be low and profit is sensitive to manufacturing costs. In this situation the major contribution of technology comes from improved production processes and management systems.

## Stage 5—decline

The mature stage may last for many years but a time comes when major changes are forced upon the industry. This can happen in two ways, both dependent upon the application of technology.

The most significant change comes with the emergence of a new technology that supplants that embodied in the traditional product. Consideration of the application of electronics, particularly in markets previously satisfied by mechanical or electromechanical products, illustrates several ways in which this can occur. The development of the pocket calculator created an entirely new market due to its ability to perform everyday calculations rapidly and cheaply. In the process it replaced the slide rule and the electromechanical desk calculator, with serious consequences for companies in those industries, in spite of the fact that the main market growth arose from new applications. In contrast, the electronic watch was a direct substitution for the mechanical watch although its low price also stimulated market growth. It also changed the nature of the market in relation to its appeal to the customer. Previously it was a technical attribute, accuracy, that provided the competitive advantage. However, with the arrival of the electronic watch all products could satisfy the purchaser's needs for accurate time-keeping. Competitive advantage had to be derived from some other customer attribute, for example visual appeal as a fashion article. In other industries, for example weighing machines, the size of the market was not affected significantly but the enhanced accuracy ensured a rapid substitution. One feature that is common to nearly all these applications of new technology is that they were exploited by companies that possessed the technical ability and applied it in markets where they may have had little or no previous experience. By contrast, the companies established in the market have often failed to build on this strength by using the new technology, which they have been unable or unwilling to adopt.

In many industries the customer impact of new technology in existing products in this stage is less dramatic. We have seen that in the mature industry the basic performance of the competing products is almost identical. A common response is to apply technology in ways that increase the in-use life of the product. This can have a serious impact on the total size of the market, which by that time is likely to be approaching saturation and dependent on replacement purchases, not because the new products are superior but because the old one is worn out. This can lead to a substantial fall in the total sales volume for the industry, with severe consequences for those engaged in it. Examples where this has occurred are stainless steel razor blades, radial ply tyres and car batteries. Today corrosion resistance is a major selling point for the motor car, reducing the main reason for car scrappage.

## Discussion

The strategic implications for technology of these stages in the evolution of an industry are summarised in Table 1. Comparison of successful with

**Table 1** Characteristics of industry life cycle evolution

| Life cycle stage | Technological emphasis | Market emphasis | Main corporate concern | Key corporate resource | Management style |
|---|---|---|---|---|---|
| Incubation | Applied research | Specialised applications | Developing a product | R & D | Entrepreneurial |
| Diversity | New product development | Development of a product | Exploiting the technology/ market, innovation | R & D | Informal, enthusiastic (early intro-duction more important than cost minimisation) |
| Sustained growth | Product improvement | Segmentation | Market share, competition | Marketing | Formalising (increasing importance of cost control) |
| Maturity | Process development | Price competition | Cost control | Production, finance | Formal (emphasis on internal efficiency) |
| Decline | Product quality (longevity), techology-based diversi-fication | Price and quality | Survival | Production, finance, technology | Formal (sometimes tending to autocratic) |

unsuccessful companies indicates that the differences in performance can often be attributed to the way in which they deployed their technology in relation to:

- the preparation for challenges from entirely new technology, a relatively infrequent occurrence but one of dramatic strategic impact when it occurs;
- new product development;
- incremental product improvement;
- new and improved manufacturing processes;
- managerial control systems.

All these areas of activity are important but the secret of success is to place the emphasis where it is likely to have the greatest impact in relation to the state of evolution of the industry. The most important strategic consideration is the recognition that there is a need for change in the nature of the technological contribution to corporate well-being.

## *Organisational implications*

The relationship between technology and organisational change is discussed in Chapter 2.1. However, it is so important that no discussion of the industry life cycle can ignore its organisational implications. These are summarised in the last two columns of Table 1, which indicate how the stage of industry evolution affects both the relative importance and status of the various functions and the style of management. It should be noted that the industry growth pattern is largely determined by factors outside the control of the individual firm, whereas the ability to respond to it is a reflection of such internal considerations as management philosophy, style and control systems. Thus the successful strategic application of technology depends upon both the identification of what should be done and the provision of an internal environment enabling it to be done—technical and organisational change are interdependent. Examination of Table 1 also provides a clue as to why it is often difficult for a large, mature company to exploit new technology. The emphasis on timeliness rather than cost minimisation and the need for a responsive, informal style of management contrasts with the increasing cost consciousness and formalisation of systems characteristic of the mature and frequently large company.

In the discussion so far it has been assumed that the industry and its core technology are following a pattern of simultaneous development. However, many companies are engaged in several industries and technologies at different stages in their life cycles. This complicates the managerial problems since it demands a diversity of technological emphasis and managerial style. This diversity is difficult for the top management of a large company to accept since there is a natural desire to have common systems throughout the company. The evolution as the industry moves through the life cycle, with its increasing degree of sophisticated managerial control, appears a natural progression. The reverse needed to exploit new technology, a degradation to the loose management characteristic of the company's early days, is associated with inefficiency. The inability to recognise the type of environment conducive to innovation is one reason why new technologies are so often exploited by entrepreneurial new venture companies.

## Technological growth and technology forecasting

Before I examine the nature of technological growth it will be useful to make a clear distinction between scientific discovery, technological progress and technological exploitation. Scientific research, which is concerned with the acquisition of knowledge of the nature of the physical world, is largely conducted without any clear objectives for its practical application; the knowledge is an end in itself. The term 'breakthrough' reflects the erratic and random nature of most scientific research. It is impossible to predict when, if ever, a particular advance will occur. The discovery of high-temperature superconductivity is a good illustration.

Technological progress, on the other hand, results from the investment of resources, often industrial, to apply scientific knowledge for a practical purpose even if the exact area of application may be uncertain in the early days of a new technology. Its aim is to establish feasibility. Technological exploitation is the embodiment of that feasibility in a product or service that meets the needs of society or individual customers.

Many years may pass before new scientific knowledge is exploited through technological progress. Sometimes this is because no immediate application can be identified and the insight of a creative person or developments in an associated and enabling technology are needed. More frequently it is not exploited because there is no strong incentive for industrial companies to invest in an uncertain activity when the current benefits of the existing technology are deemed adequate by the organisations that might adopt it. That situation will not last for ever and eventually a time will come when competitive pressures stimulate a number of companies to invest in the new technology at about the same time. It is important to recognise the causality of this process. Resources will not be made available for the new technology unless a financial benefit derived from the market is anticipated; thus technological progress is largely market-driven. Furthermore, because a number of organisations are usually involved, this process can be regarded as exogenous to the individual firm. Examination of the path of progress of a large number of technologies shows that this results in growth which follows a regular pattern. This regularity makes it possible to use the pattern as a model for forecasting the future rate of advance as a basis for decision-making. This is the rationale behind technology forecasting.

An individual company may contribute to the advance of a technology but because of the number of companies involved in most technologies the action of any one of them is unlikely to have more than a marginal effect upon the rate of progress. A company's concern is to incorporate the technology in a specific application to serve an identified consumer need. This is a conscious act of investment in what may become an expensive development project. This decision has to be taken in relation to the company's business and market strategies. Furthermore, it may be only one of a number of technologies embodied in the end-product, which must be considered as a total system.

Few companies, except in the new science-based industries, engage in scientific research. A greater number are involved in longer-range applied research to develop their knowledge base. However, neither of these activities can be justified except in terms of their eventual exploitation in marketable products that will contribute to corporate growth and profitability.

## Patterns of technological progress

The typical pattern for the path of technological progress is an S-curve (Figure 4). This curve is usually drawn with a technical parameter and time as the axes. From the previous discussion it has been seen that the actual causal relationship is between market needs and financial investment. This is important to bear in mind when examining an S-curve for two reasons.

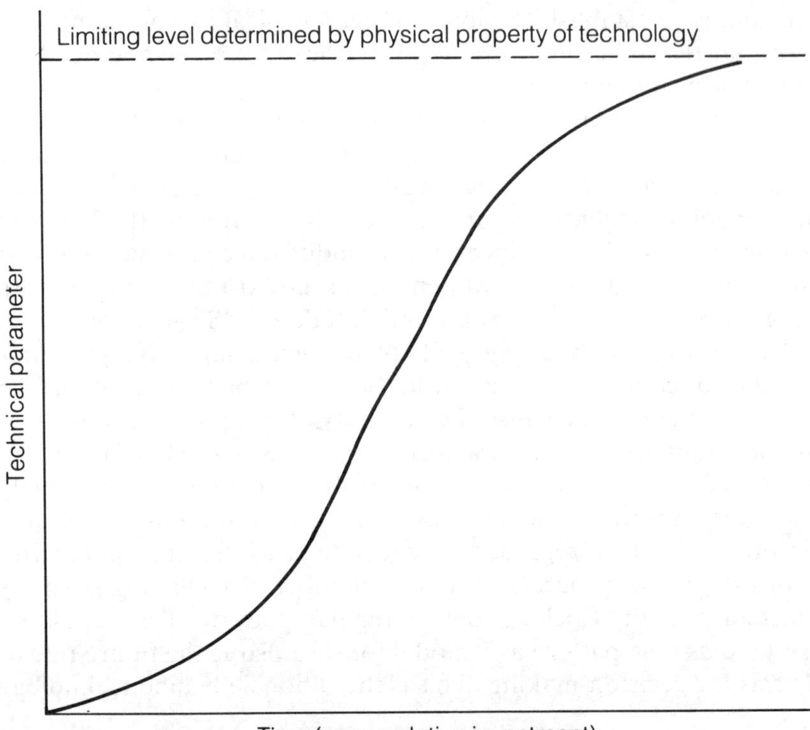


**Figure 4** The S-curve of technical progress.

First, it would be more appropriate to plot cumulative investment in the technology rather than time on the *x*-axis. However, for most technologies it is not possible to establish a figure for the growth of cumulative investment so time is used as a surrogate. Although this is satisfactory for most purposes it must be appreciated that the intervention of some new factor may cause a curtailment of investment and hence the progress of the technology.

Second, careful attention must be paid to the selection of the appropriate technical parameter. It must reflect the needs of the attribute the market needs from the technology, which is not necessarily what is uppermost in the mind of the technologist. A classic example of this is provided by the emergence of the jet engine in civil air transport. The attribute required by the user is economy reflected in the parameter, cost per passenger mile, a function of the total technical system. This was overlooked by some aero-engine manufacturers, who dismissed the potential of the commercial jet aircraft because of the high specific fuel consumption of the engine, the parameter which hitherto had been their main concern. It must also be appreciated that in some technologies a stage is reached when market needs are adequately satisfied although the full potential has not been realised, for example watch accuracy or car corrosion resistance.

If these two provisos are borne in mind the S-curve provides a valuable tool for forecasting future technological progress, because if no reason can be identified to curtail that progress, there is a strong presumption that the past growth will be maintained into the future. Examination of the S-curves for many technologies might suggest that the progress of a technology is inevitable. This is not so, for it must be stressed that the stimulus for technology must come from the market, which alone will determine whether the resources to maintain the progress will be made available.

The reader will have noted that the S-curve of technological progress is of similar shape to the industrial life cycle discussed earlier. There is, however, an important difference. The industrial life cycle is a conceptual diagram that helps understanding of the pattern of evolution but it cannot be drawn accurately: it relates to a broad category of products. The S-curve, in contrast, is specific to a particular technology and can be plotted within the accuracy of the available data and used as a basis for quantitative forecasting. The industry life cycle represents the application of one or more technologies in marketed products whereas the S-curve represents the past and future level of technological performance that is a potential for the development of products. It is important that all technologists identify the attributes required by the market, relate these to technological parameters and quantify them.

The evolution of the S-curve can be conveniently considered as occurring in three stages. In the initial period of slow growth many major uncertainties remain, the solution of which requires time and considerable resources. The period of rapid advance that follows leads to the ability to exploit the technology in a range of applications of increasing performance. This may not necessarily coincide with the period of rapid growth for the industry based upon it. In the third stage progress slows as it approaches the limiting level determined by the physical nature of the technology itself; this level cannot be breached using that technology. It can be seen that the investment to achieve an incremental improvement in performance escalates as the limit is approached. In other words the productivity of R & D expressed as the improvement in performance for a given expenditure falls rapidly.

The procedure for using the S-curve to aid technical decision-making is as follows:

1. Identify the attribute of the product desired by the market.
2. Determine the technical parameter by which this attribute can be measured.
3. Establish the upper limit of this parameter as determined by the physical properties of the technology.
4. Plot the past performance of this parameter over time, i.e. the progress along the S-curve to date.
5. Fit an S-curve (logistic) to these data and extrapolate into the future.
6. Use the extrapolated curve to establish the forecasted performance at dates in the future relevant to product development.
7. Question whether there are any factors that might inhibit the forecasted growth.

### *Technological substitution*

Although the limit for any technology is set by its physical properties, there may be another technology, usually new, which has a higher limit for the same parameter (Figure 5). The history of technology shows that this is often the case. For example, the speeds of operation of many mechanisms, such as cams, were approaching their limits due to inertial effects when this restriction was removed by the advances in microelectronic systems. In future the higher operating temperatures of ceramics are likely to sustain the advance of internal combustion engine performance beyond that achievable with metals. Thus product performance continues to improve as a result of the application of a sequence of different technologies, each with a higher limiting level for the technical parameter.

The competitive threat from the new technology is obviously greatest when the progress of the old is slowing. There is, of course, no theoretical reason why the new technology cannot emerge at some other time. However, this is rarely

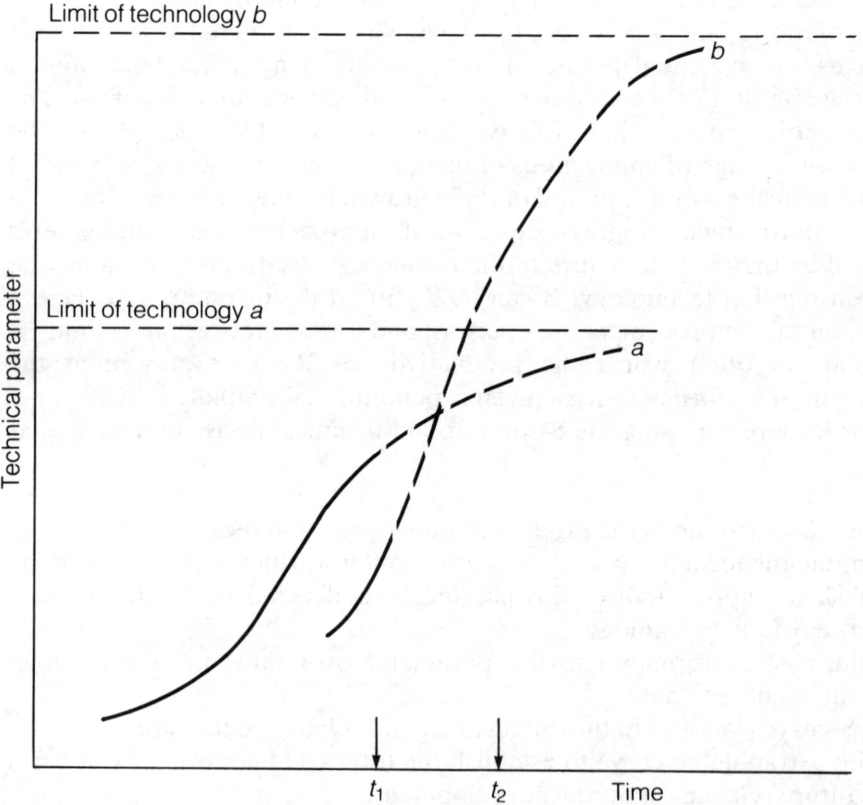


**Figure 5** Technological substitution. At time $t_1$ the performance of technology *a* is inferior to that of technology *b*, *but* examination of the curves indicates that by time $t_2$ the performance of technology *b* exceeds the maximum achievable by technology *a*.

the case and the reason can probably be attributed to the fact that there is no great incentive to invest in a new technology that may still be advancing slowly and faced with technical uncertainties while the performance of the existing technology still has a considerable development potential.

The important implication for the technologist is that he or she must identify the emerging technology, plot its development to date and compare it with the position of present technology on its S-curve. Only thus can the relative significance of the two technologies be evaluated. The greatest danger (or opportunity) occurs when the new technology reaches the stage of rapid growth. The impact of a new technology has often been ignored because its performance in its early stages may be inferior to the established technology. The value of considering technologies in relation to their S-curves is to direct attention to the future, which is likely to identify the increased potential of the new. Where there is a higher limit it can be confidently predicted that there is a high probability that it will eventually replace the existing technology, as no amount of investment can take the technology beyond its natural limit. The ability to forecast these developments enables a change in the R & D strategy to be made in sufficient time to meet the threat and to build up resources in the new technology.

## *Attribute substitution*

Customers do not buy technologies, they buy products or services that satisfy their needs. These needs are complex and comprise a number of attributes, each depending upon the contribution of one or more technologies. In the motor car, for example, they would include speed, acceleration, fuel economy, road holding, reliability, comfort and so on. To some extent the mix of attribute performance in the product depends upon the market segment, which is a corporate rather than a technical decision. However, within each segment there will be technical decisions about the development effort to be devoted to each of the attributes. In some cases there will be trade-offs between attributes, for example speed and fuel economy; in other cases they will be independent of each other, for example comfort and reliability.

Each of these attributes can be related to a technical parameter which will be at some position on its S-curve. Comparison of the curves for the various parameters (Figure 6) is likely to show that the productivity of R & D for some attributes is higher than for others. For some the productivity will be increasing and for others decreasing; the latter are likely to include development areas that have been stressed in the past. Thus, examination of the various curves relevant to a product might reveal that a shift of emphasis would yield a greater competitive advantage for a given expenditure than can be obtained from the current R & D focus. These changes in emphasis will usually occur within the industry as a consequence of competitive forces. The value of carrying out a formal analysis is that it enables the factors to be examined explicitly and provides a systematic basis for anticipating the changes that will occur. This is

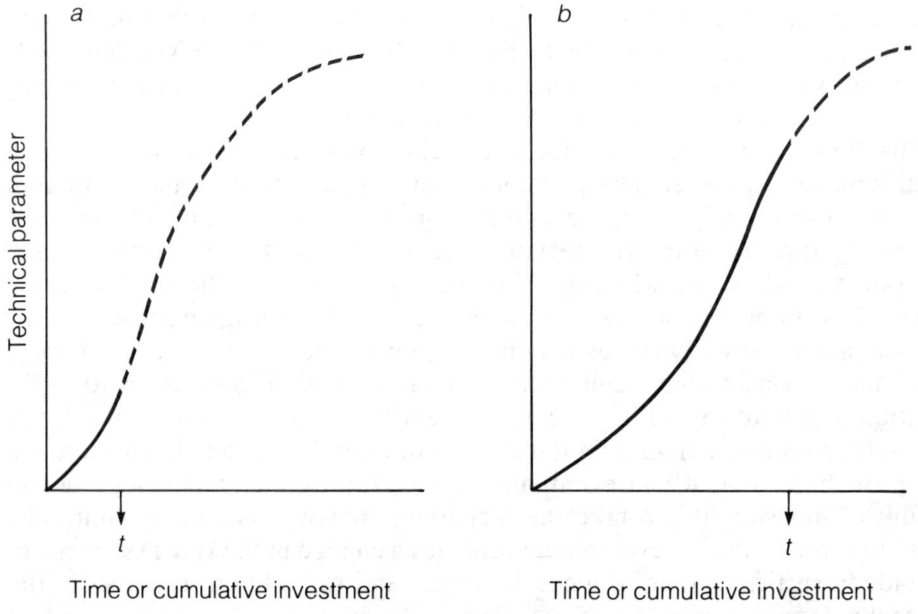


**Figure 6** Attribute substitution for a product satisfying more than one market attribute. At time *t* attribute *a* has the potential for considerable performance improvement. At time *t* attribute *b* is approaching its limiting level. Increasing investment is necessary to achieve marginal performance improvement.

the main justification for the use of all the forecasting techniques. The study of the development of technology using forecasting methods provides a competitive advantage through the ability to adopt a proactive technological strategy.

There are also occasions when the relative importance of the attributes to the customer will change. As we have seen earlier a performance may be achieved that fully meets the needs of the customer; further technical improvement is then of little value. Sometimes external factors such as an increase in fuel price change the customer's priorities. Sensitivity to these influences ensures that the technical effort is deployed in a way that contributes most to the needs of the purchaser of the product, thereby enhancing the company's competitiveness. Many companies fail to identify when change is necessary and consequently do not obtain the best return from their investment in technology. These changes may require the adoption of a technology new to the company or indeed may not be feasible without the emergence of an entirely new technology.

Similar considerations can be applied to a technical system that is an integrated combination of interrelated components. It is possible to evaluate which part of the system can contribute most to the performance enhancement of the total system at minimum cost. This underlines the argument that runs through this chapter: the corporate contribution of technology depends as much upon how the resources are deployed as upon the size of the resources.

## *Product substitution*

The analysis of the S-curve, and the technological and attribute substitutions, aid the technical decision-maker in assessment of the level of technological performance likely to be reached in the future, in identification and evaluation of emergent technologies, and in determination of the product attributes that are likely to represent the most productive deployment of technical resources. These approaches are most valuable for companies that aim for technical leadership since they give a good indication of what is achievable and beneficial; in other words for companies with an offensive strategy.

If a defensive strategy is adopted the company does not aim to be first into the market with the new technology. There must be recognition that it must be adopted at some time but the key question to be addressed is: when should they introduce products incorporating the new technology? Interest is focused on the rate at which products based on it will diffuse into the economy and replace the old technology.

Analysis of a large number of substitutions reveals that once again there are identifiable patterns, also based upon a curve of S shape. Examples of this include detergents for soap, colour for monochrome TV, radial for cross-ply tyres and diesel for petrol-engined motor cars. A technique developed by Fisher and Pry (1971) is one of the most useful methodologies available to the business and technical strategist (Figure 7). This is based upon the observation that once the new technology has gained about 5% of the market it is highly probable that it will ultimately achieve full substitution and that the dynamics of this early part of the substitution provide an adequate basis for forecasting the total process. It is in effect a technique for long-term market forecasting.

Before applying the technique it is necessary to gain an understanding of the impact of the new technology, as it does not always manifest itself as a straight substitution. Earlier we mentioned the pocket calculator, which created an entirely new market and could not be regarded solely as a substitution for the slide rule. Another case is the microwave oven, which was adopted as an addition to rather than as a replacement of existing means of cooking. In both cases, however, the diffusion followed a curve of S shape. Sometimes the effect of a new technology is to segment a previously integrated market where some applications remain satisfied by the old technology and others are substituted; examples of this are soap and detergent, and synthetic and natural fibres. In such cases it is necessary to assess the relative sizes of the two segments and then to apply the substitution analysis only to that segment being affected. These provisos illustrate a general point in relation to the use of forecasting techniques. While the technique can provide a useful aid for the decision-maker it must not be applied mechanistically without an informed understanding of the total process and a realistic assessment of the impact of the technology on the market.

Figure 7*a* shows the shapes of the growth curve for the new technology and of the decay of the old. When the ratio of the new to the old is plotted on log-linear graph paper a straight line is obtained, which can be extrapolated to

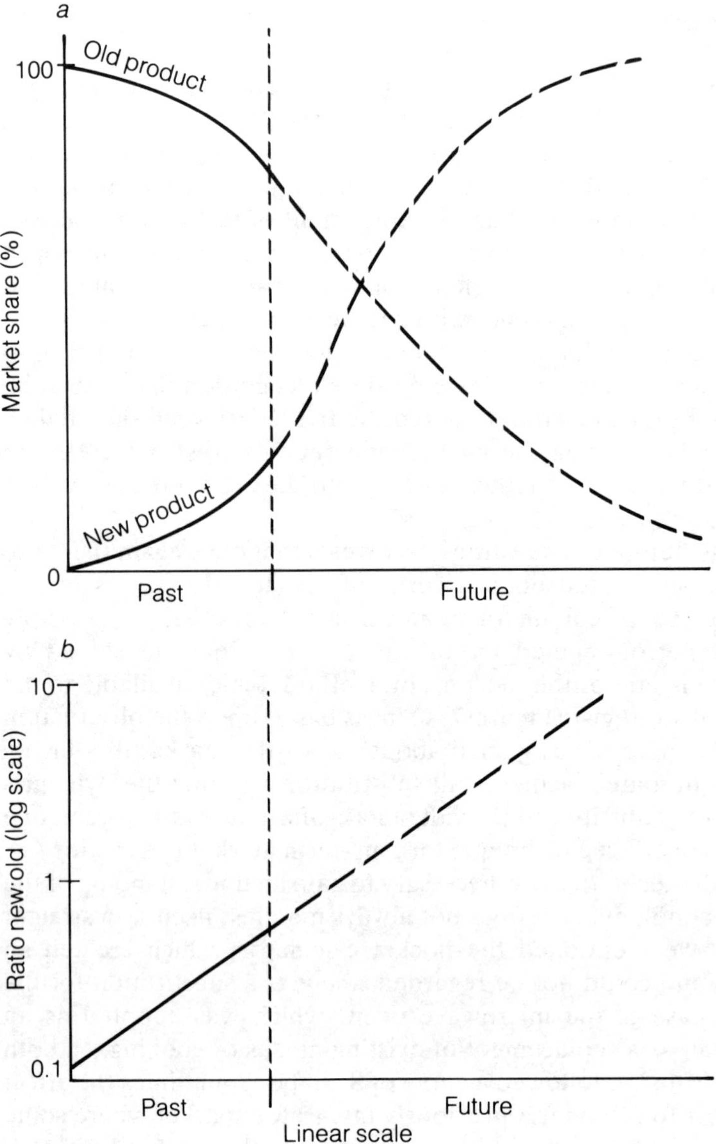


**Figure 7** Product substitution (from Fisher and Pry, 1976).

forecast this ratio in the future (Figure 7*b*). This is one of the most useful of the forecasting techniques applicable over a wide range of technology-based product substitutions. It draws attention to two periods when companies often misjudge the rate of market growth, namely the onset of the rapid rate of substitution once the new technology has established itself and the slowing as market saturation is approached.

## *Technology forecasting techniques*

It is beyond the scope of this chapter to describe the many techniques of technology forecasting. They fall into two broad categories—extrapolative and normative. The extrapolative techniques are mostly based upon the S-curve and its many applications already briefly discussed. The normative techniques are directed towards the long-term objectives, where the means of achieving them are not clearly understood and where quantitative objective data do not exist or are only partially available. Among the most frequently used of these techniques are Delphi, scenarios, morphological analysis, cross-impact analysis, modelling and relevance trees.

In all cases the purpose of technology forecasting is to:

- identify new technologies and evaluate their significance;
- relate a measure of technical performance to a time scale;
- associate the forecast with a probability, since no forecast can be expected to yield a firm prediction of the future;
- provide a systematic analysis to support an informed discussion on technological developments as a basis for decision-making.

## Conclusion

The main rationale for developing a business strategy is the need for change. If that were not the case the company could continue with its traditional activities; success would come from doing more of the same thing rather than doing different things. The stimulus for change comes from the threats and opportunities arising from the environment in which the company operates and the actions of competitors. Technology is one of the most important of these stimuli but cannot be isolated from other elements of the business environment. The process of strategy formulation, be it corporate or technological, consists of a systematic analysis of a complex web of factors and their interactions in relation to an explicit set of objectives. It is translated into practice through the allocation of resources.

It has been said that 'strategy is simple but not easy'. The concepts described in this chapter are not intellectually demanding. However, the problem arises in applying them to an evolving situation where there are a number of conflicting interests to be reconciled—financial, marketing, technological. The corporate level is concerned with a portfolio of businesses, marketing with a portfolio of products, and technology with a portfolio of disciplines. One must also consider the balance between short-term interests, likely to be more incremental in nature, and the long term, which may require the exploitation of radical new technologies. It must also be recognised that investment in new technology usually involves a higher degree of risk, at least in the short term.

There are no simple methods for reconciling these differences. In the final analysis reconciliation must be based upon judgement, but informed judgement founded on a systematic analysis of all the influences that bear upon the future of the business. It is essential that technological considerations are fully represented. This manifests itself most strongly in two ways. First, it is only the technologist who is likely to appreciate fully the potential of a new technology. Second, long-term success depends upon the health of the corporate technology in developing its knowledge base within a coherent technological strategy that is likely to have a long-term orientation. The furtherance of these objectives must be a major concern of the technologist although he or she must recognise that they may need to be modified in the light of the total corporate situation.

In this chapter I have examined a number of considerations affecting the technological strategy formulation process. In summary they are:

- The need to determine priorities for the allocation of scarce technical resources between conflicting demands.
- The evolution of the industry life cycle and its influence on the appropriate balance between these demands.
- The identification and evaluation of the strategic and market potential of new technologies.
- The assessment of the time scale of technical progress and the appropriate time for investing in a new technology aided by technology forecasting.

The discussion has focused on identifying what might and should be done, which is largely an intellectual exercise, manifesting itself as a change within the business and technological strategies. It must be recognised, however, that this is only the first step in a complex process involving possible changes to the corporate culture, the organisation, managerial style and corporate systems, without which the implementation of technical change is unlikely to achieve its full potential.

## Reference

Fisher, J. C. and Pry, R. H. (1971). A simple substitution model of technological change. *Technological Forecasting and Social Change*, **3**.

## Further Reading

Foster, R. (1986). *Innovation: The Attacker's Advantage*. London: Macmillan.
Goodridge, R. and Twiss, B. C. (1986). *Management Development and Technological Innovation in Japan*. London: Manpower Services Commission.
Linstone, H. A. and Sahal, D. (1976). *Technological Substitution*. Amsterdam: Elsevier.

Martino, J. P. (1983). *Technological Forecasting for Decision Making*, 2nd edition. Amsterdam: Elsevier.
Sahal, D. (1981). *Patterns of Technological Innovation*. Wokingham: Addison-Wesley.
Twiss, B. C. (1986). *Managing Technological Innovation*, 3rd edition. London: Pitman.
Twiss, B. C. and Goodridge, M. (1989). *Managing Technology for Competitive Advantage*. London: Pitman.

# 1.2 The Management of Research and Development

Alan W. Pearson

The level of expenditure on activities that come under the heading of research and development varies between industries and between companies in the same industry. At the lower end less than 1% of turnover is allocated to it, while at the extreme—for example in fast-growing high-technology companies—the figure could be as high as 50% for short periods of time. The relative importance placed on product or process development, the emphasis placed on short- or long-term needs, on major innovations or minor improvements, on being first in the field or a follower, and even upon the choice of having centralised or decentralised facilities, have an impact upon the size and nature of expenditure on R & D.

Differences arise from the basic characteristics of the industry, for example from factors such as its maturity, and in particular the maturity of the technologies on which it is based. The way in which an organisation chooses to use R & D is also important, including its choice of whether to do it all in-house or to buy-in significant quantities as contract research, and whether, for example, it places more emphasis on licensing or acquisition policies than on ones based on internally generated developments.

It is clear therefore that those concerned with the management of R & D must pay attention to the management of the business. This is not to suggest that their actions must be dominated by the business activity, but rather that they should be seen as clearly relevant to the overall objectives. R & D's function is to make sure that the business can make the best use of available science and technology to develop and maintain competitive advantage. There is rarely any dispute with a statement such as this, but differences in practice can, and do, arise owing to the time horizons within which the different functional areas of a business are often forced to operate.

The management of R & D must be concerned with setting and agreeing objectives that are consistent with those of the business, with building up a resource base in the skill areas necessary to support those objectives, and with providing an organisational structure that ensures that these resources are used to progress projects and programmes of work that contribute effectively to short- as well as long-term needs. However, it must be remembered that R & D only provides information, and if this is not translated effectively into exploitable outputs it is of little value. The evidence is that if the work is relevant and the output well used the return on investment in R & D far exceeds that from other areas of the business.

In this chapter, therefore, I will look at those areas that impact directly on the effectiveness of R & D. I shall pick up the topic discussed in Chapter 1.1, and will look here at the relationship between R & D and corporate policy. I will also anticipate some of the discussions in Part 2 of the book in looking at organisation structures for R & D, and will then concentrate on project evaluation and selection, planning and monitoring, and performance audit.

## Relating R & D to the Business

An organisation must be able to recognise the opportunities that science and technology can open up, as well as the threats that science and technology can pose if used by other organisations to gain competitive advantage. It is essential to have strong links between the various business units and R & D, which forces attention on both the short and the long term. A simple form of presentation that brings out the importance of this dialogue is the relevance tree. An example of this is shown in Figure 1.

This type of diagram makes it clear that the ability to carry a project through to completion depends upon the quality of the technology available in the organisation, which in turn depends upon the background or basic research in relevant disciplines. In general, scientists are recruited for their expertise in these disciplines and not specifically to satisfy the demands of the current workload. However, the projects that can be undertaken at any given time depend upon the type and quality of expertise available, and it must be

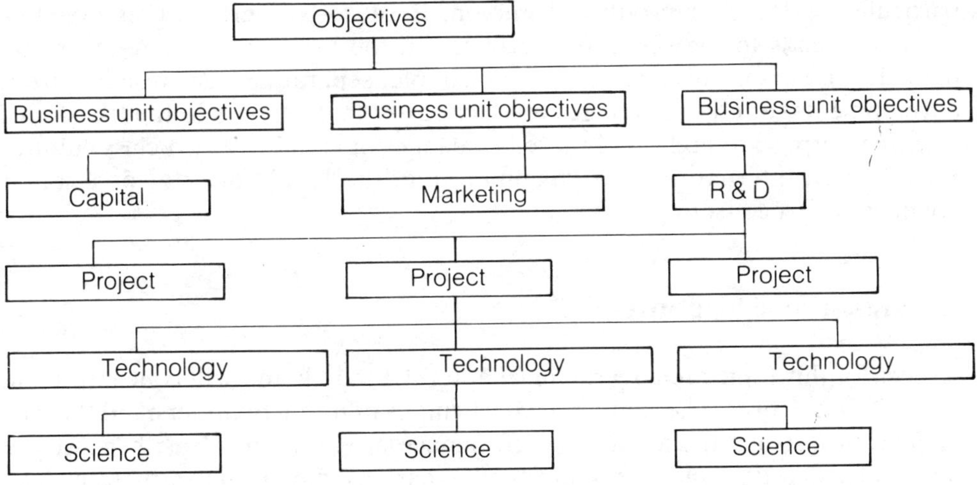


**Figure 1** An example relevance tree.

remembered that any major change in direction involving new or different technologies will almost certainly require a change in the resource base.

This does not happen instantly. Scientists and technologists cannot be hired and fired at short notice, and a major objective of R & D must therefore be to develop a skill basis relevant to the future needs of the business. This means that individual business units must be aware of their likely longer-term needs as well as the more immediate concerns they have in striving to achieve their objectives. The relevance tree emphasises the need for a constructive dialogue between each business unit and R & D on a regular basis. In such discussions factors to be taken into account will include the changing external environment, social and political forces, competitive pressures, the likely impact of new and emerging technologies, product and technology cycles, S-curves, etc.

Unfortunately it is often the case in practice that too much emphasis is placed on meeting today's needs and not enough concern is shown for the longer term. In many organisations projects are paid for directly, i.e. a customer–contractor type of relationship is adopted, and there are some clear advantages in relating financial support to work done because of the closer working relationship, communication and commitment that follow. More flexibility is gained to mutual benefit when business units are encouraged to negotiate a longer-term overall commitment to R & D for work within certain boundaries that are seen as relevant to their interests. Corporate support to R & D may also be necessary, possibly associated with the more basic and background work in key areas of science and technology that might be of relevance to a number of business units in the future, or that might even stimulate developments leading to the setting up of new business units.

Considerable care needs to be given to the type of financial arrangement used, and in particular attention needs to be paid to the way in which any funds allocated to basic research are used. The evidence in practice is that if care is not taken these are not well applied, mainly because the work does not have a particular 'sponsor' or product champion. Because of overload pressures this inevitably leads to the short term driving out the long term. There are many possible answers to this problem—for example, separating basic research from development—but disadvantages often cited are that barriers to the transfer of research output are high, and there is a danger of creating a two-class culture. Providing an internal product champion, e.g. the R & D director or a senior scientist, can be useful.

## Organisation Structure

Aspects of this topic are covered in Chapter 2.1, and here I will concentrate on the R & D function. There are many difficulties in trying to understand the way organisation structure can affect performance in R & D, not least because the informal can often play as important a part as the formal structure. The importance in R & D of earned rather than given authority must not be underestimated, and it is well established that the architecture of the building can influence communication patterns as much as the organisation structure. In practice a variety of structures may be present within the same organisation. We can consider three types—the functional, the project and the matrix. The functional is perhaps best described as a structure in which the emphasis is on hierarchical structure based on a single function or discipline. The project structure, as the name implies, emphasises a focus on the task to be done at a particular point in time, whether it be short- or long-term. The matrix is a structure characterised by the fact that people are essentially a part of a discipline-based hierarchy (which is the one to which they relate for rewards and promotions, and which is the one said to be responsible for ensuring that technical quality standards are monitored) and also part of one or more project teams (which are formed to ensure that work is oriented towards customer needs, and in particular is carried out in such a way as to achieve agreed cost and time standards). Chapter 2.1 has further details of these structures.

To achieve and maintain technical quality requires critical mass in any given area, good contact with peers inside and outside the organisation, and a promotion system based on technical competence, often in a very limited area and not necessarily required immediately within the organisation. Being part of a project team seems to be the opposite, with short-term needs tending to drive out longer-term personal development. The matrix, with all its potential shortcomings, is seen as offering advantages over both these structures.

In practice, structures tend to be more fluid than these three distinctions might suggest, and other combinations have been described, for example the project matrix, balanced matrix and functional matrix. It should also be noted that 'project manager' can be a transient or permanent position, low or high in the hierarchy. It has been shown that technical quality can be maintained in a project matrix type of structure, and also that the same types of structure are used in different ways by different organisations. For example, both functional and project structures have been found in basic research laboratories, and matrix and project structures are used in product development areas. An explanation for this can be found in such factors as the nature of the technology, the size of the projects, and the quality and experience of the scientists and project managers.

This has led to more concern being given to the actual management of the work, and to the planning, control and resource allocation process. This is not surprising, because it is clear that even the project structure cannot operate well if there are inadequate resources available, people are not committed to the goals, and they are not capable of or motivated towards collaborating with their colleagues. In a matrix structure, with people working on more than one project at a time, team building and performance development of all individuals become very important. More attention is therefore being paid to this area, and findings from the behavioural sciences more extensively applied. Emphasis is placed on leadership style, participation, team-building, establishing and clarifying goals, roles and procedures, and the importance of improving communication, negotiation and conflict resolution skills.

All of the above apply equally well to teams that cross functional boundaries, and real value is obtained when all parties realise their differing needs and the importance of negotiated trade-offs at key decision points. For many people this is a fairly recent change, and the realisation that such skills are important is having an impact upon management development programmes for scientists and technologists. This trend is not only in line with organisation needs, but fits well with the now accepted fact that a key motivator in R & D is work challenge, and that working in teams and contributing to successful outcomes produces great satisfaction. What may be required, therefore, is less emphasis on and concern about organisational structure and more about providing adequate resources and a supportive environment within which people feel they can make a significant contribution. This is clearly an issue that can be debated, but in a time of flattening structures, lower growth and fewer promotional prospects, good project leadership plays an important part in the success or otherwise of any organisation.

## Project Evaluation and Selection

It would be very easy to argue that R & D projects should be evaluated using the types of criteria that are used in many other areas of the organisation, such as return on investment using the net present value or internal rate of return methods. The requirement for such approaches to be used is that reasonable estimates can be made of the expected cash flows over time. In some areas, for example process improvement, this should be possible, although even here research suggests the errors can be extremely large. For a large proportion of R & D activity the evidence is that estimates are definitely not reliable enough to allow such methods to be used. In particular, the estimates of benefits are very difficult to obtain because of the influence of so many variables. They are not only dependent upon the level of technical performance achieved, but also upon the timing, competitive action and reaction, strategies used for exploitation and marketing support allocated. The lack of such data has meant that other, more sophisticated techniques, for example linear programming, have proved to be of less value than was originally hoped, despite a considerable amount of research effort. Their major value has been said to be the emphasis they place on the importance of a balanced portfolio. The best one can say about the more quantitative techniques therefore is that they should be used wherever possible, but that their limitations should be appreciated. In particular, any attempt to force data to fit the method is likely to prove to be of little value. Alternative approaches must therefore be looked for, and the ones most commonly encountered in practice include various forms of checklist.

Checklists are relatively easy to use. As the name implies, they are essentially a list of questions that must be answered for any particular project. These are often very simple. For example, is the project technically feasible? Do we have the resources to manage it effectively? Is there a real customer

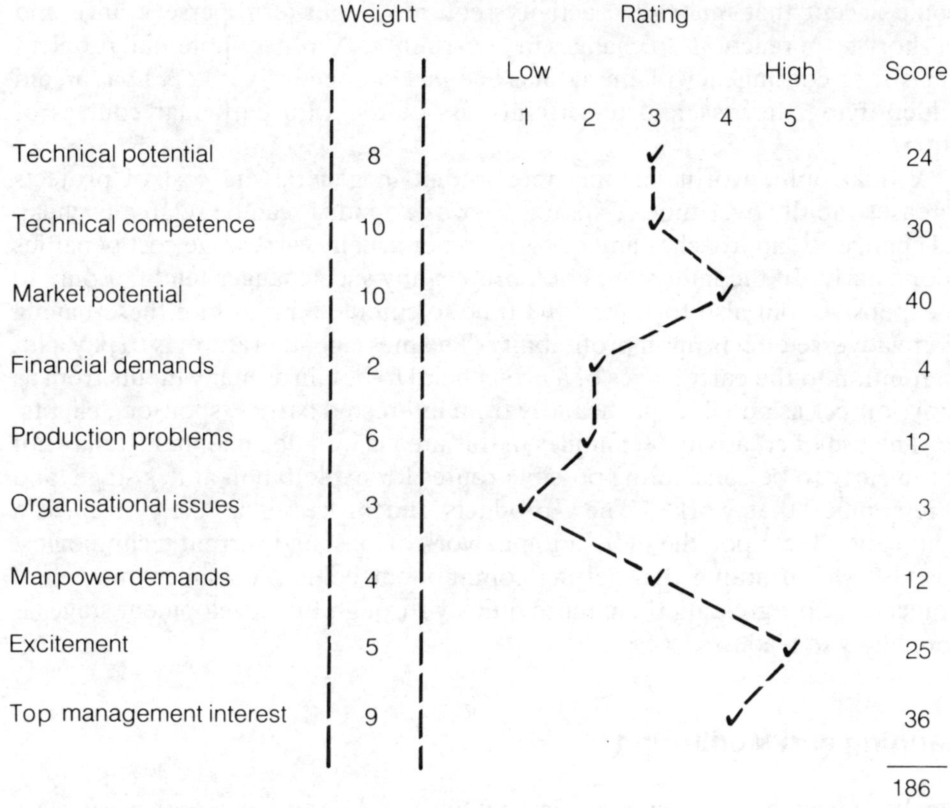


**Figure 2** Example graded and weighted checklist.

need? The checklist can be extended by asking for answers to be given on a graded scale and further elaborated by weighting the various questions in order of perceived importance. An example is shown in Figure 2.

Many people have emphasised the importance of identifying criteria before attempting to evaluate the alternatives, and then making use of pairwise comparison approaches. They do not seem to have been used to any great extent until the relatively recent development of microcomputer-based models based on the analytic hierarchy process (AHP) methodology. The potential for inclusion of both qualitative and quantitative data with this approach has been demonstrated. Early experience with these models in R & D suggests they are very promising, not least because they really do promote effective communication and enable people from different functional areas to make significant contributions to the decision-making process.

When used properly, all these methods, from the simple checklist to the more sophisticated computer-based models, can encourage creativity and idea generation. The result is that project selection is seen not as a simple accept/reject decision, but as a process that encourages viewpoints that improve proposals and even spawn new ones. Additional value is obtained when the same process is applied to ongoing projects, provided, of course, that people accept that innovative activity requires longer-term perseverance and not short-term reactions to changes in uncertainty. A form of potential problem analysis, or contingency planning, has been used effectively in R & D as an aid to identifying the risk and uncertainty associated with particular courses of action.

A final point worth making here is that in general the cost of projects increases rapidly over time. Exploratory work on ideas can be relatively cheap and changes to approaches and designs do not usually carry large cost penalties if done early. In the later stages of a project any such changes tend not only to be expensive, but also to extend the time to completion, both of these having severe adverse effects upon profitability. The message, therefore, is to pay a lot of attention to the early stages of a project and to obtain as many inputs from as many sources as possible, particularly from interested parties, sponsors, clients, etc. The use of creativity techniques in this area is well documented. It may not be too early to be considering possible routes for exploitation at this stage, and to remember that work on new products and processes is likely to have a significant effect upon the motivation to work on past and current technologies. It is also worth noting that getting commitment behind a smaller number of projects and progressing them more quickly through the development stages is more likely to lead to success.

## Planning and Monitoring

The literature on project management in R & D contains some very clear messages, related specifically to the barriers to success. Any list of these would almost certainly include unclear objectives, insufficient resources, shifting priorities, failure to anticipate potential problems, lack of milestones and progress reviews, lack of commitment, and failing to learn from experience. The aim must therefore be to design a simple planning and monitoring system that will help in tackling these issues, bearing in mind that the more detailed management of projects must get down to face-to-face discussion between people who are knowledgeable about the technical content.

In considering this challenge there is no shortage of techniques that might be of use. An example that is well known in R & D is the bar or Gantt chart, with its focus on identifying key activities and highlighting the resource requirements for each, including planned start and finish dates. It is not difficult to couple this to a responsibility chart and an organisational chart, if so desired, and to produce as an output a plan of the resource requirements over time. These can be checked against the available resources, taking into account other projects already ongoing or in the pipeline. A number of computer programs, mainframe- and microcomputer-based, are available to assist in the presentation and analysis of such information.

The next most well-known technique, again much in evidence in parts of R & D, is the critical path type of network. This can be based on the activity-on-arrow or activity-on-node procedures. It is well described in the

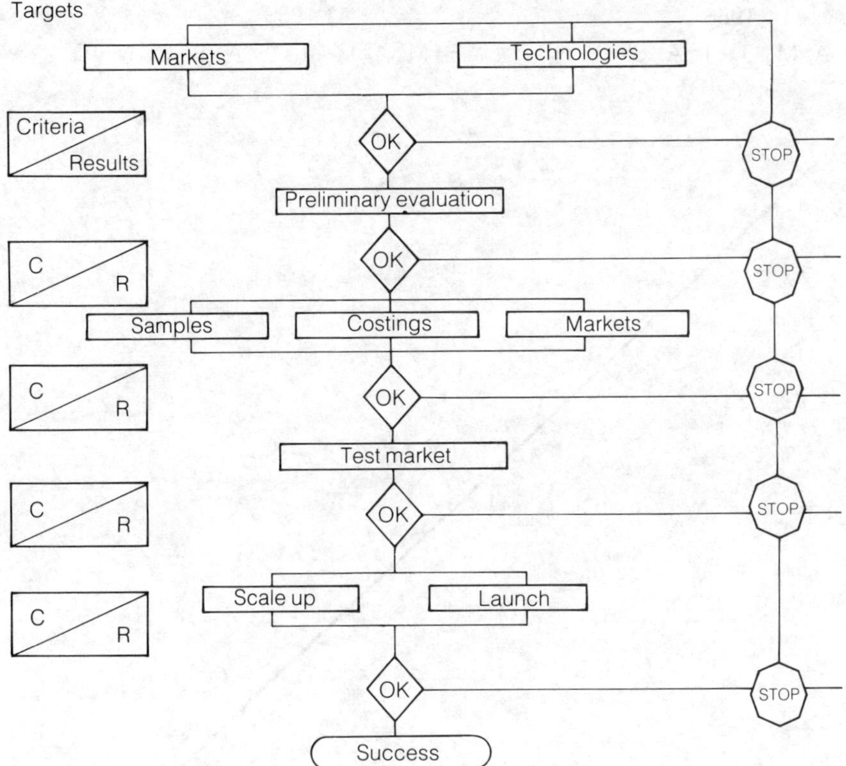


**Figure 3** Example flow diagram, or research planning diagram.

literature and packages are available on the mainframe and microcomputer. An advantage over the bar chart is the ability to show the interrelationship between activities, from which it is not difficult to calculate the critical path and the shortest finishing time for the project. The bar chart is commonly used as an addition or output from the network for resource management purposes. In the network method uncertainty can be taken into account through the use of the three-estimate approach, i.e. optimistic, most likely, and pessimistic, with the expected value being calculated as a weighted average. It is possible to allow for certain types of uncertainty by introducing branching nodes to allow the recycling of activities, but this variation is not often seen in practice in R & D.

A useful variation of the activity-on-node approach is now commonly seen in the form of a flow diagram, sometimes referred to as a research planning diagram (RPD), as shown in Figure 3. Such a diagram includes the addition of decision points at which specific questions are usually asked. The answers to these questions indicate in which direction further work is likely to go, including the possibility that the project should be stopped.

These diagrams, with their successive questioning and multiple possible outcomes, are particularly suited to many areas of R & D, highlighting as they do the inherent uncertainty of the work. Maximum benefit is obtained when they are linked to a simple monitoring procedure which indicates the current

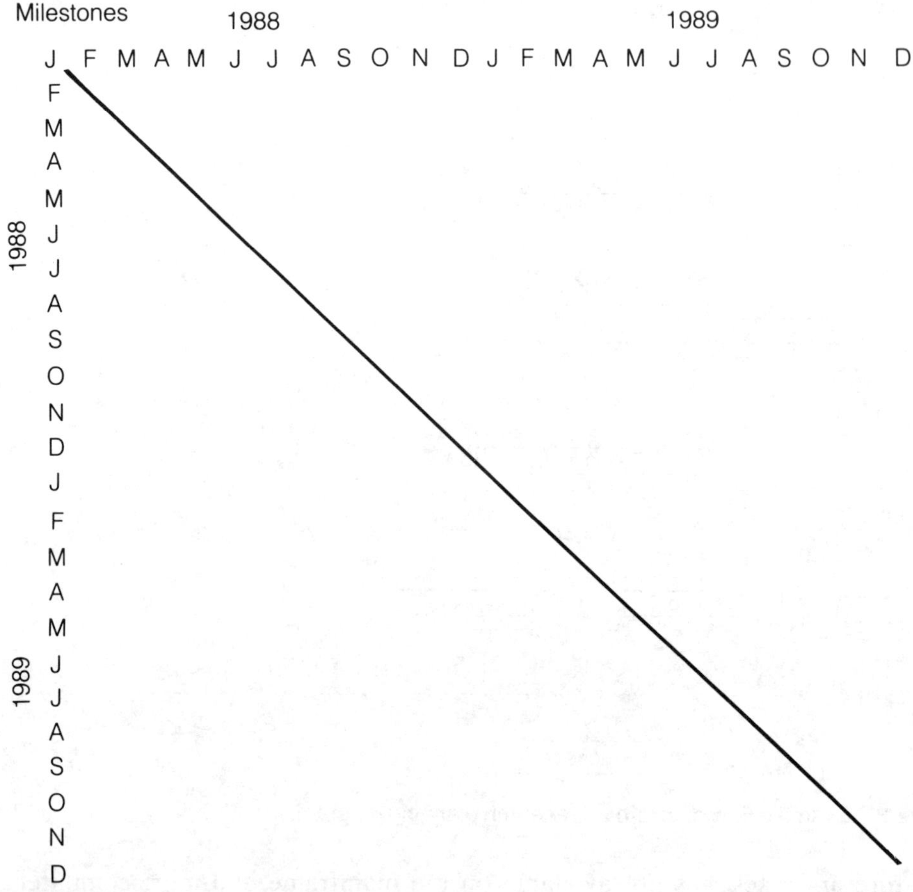


**Figure 4** Example time-based progress chart.

and projected state of progress. A procedure that has been found to be most useful in practice focuses attention on milestones. These are easily identifiable from key decision points on the flow diagram, but the value of the milestone monitoring approach is that this same information can be derived from many other forms of planning technique, e.g. bar charts and networks. Milestones can also be identified without a detailed form of plan, although this is not generally a desirable thing to do. Whatever the starting point, it is possible to develop a system that is very flexible and simple to operate in practice, based on the slip or progress chart method.

Progress charts take various forms, and they can be either time- or cost-based. Time-based ones are generally found to be the simplest to interpret and the most useful in practice. They portray calendar and review time on the same diagram, and one variation is shown in Figure 4.

Progress is reported by adding to the diagram updated forecasts of the time required to meet the next and all subsequent milestones for which relevant information is available. They therefore form the basis for a forward, and not simply historical, monitoring system. Times are revised in the light not only of technical progress, but also of any foreseen changes in goals, priorities, resource availability, etc. Additional information can usefully be gained about progress if reasons are given for any anticipated changes. Statements of actions that might be taken to correct foreseen slippage can also be added to the information provided. Potential problems and contingency plans are therefore highlighted.

Cost/progress charts more often use ratios devised from information provided about the original estimate, the cost to date and the estimated cost to completion. An example is shown in Figure 5.

The time/progress chart can also be put into ratio form or the cost/progress chart can be put on to a time basis to suit the preference of the users. The advantage of a common format is essentially that useful management information can be gained from an analysis of both the cost and time progress. It must be emphasised that this is the prime purpose of a good planning and monitoring system—to give useful information on which action can be taken in time to improve the performance of a project or to close one down as quickly as possible if this is what is indicated. Stopping projects has always been seen as a more difficult task than starting them, because of issues such as sunk costs, momentum, motivation, etc. But stopping projects earlier rather than later not only saves resources, it also has a real opportunity benefit in that these same resources can be transferred to areas where the outcomes are likely to be more successful. In the long term this will have a very positive effect not only on profits but also on motivation.

The benefits of what is essentially a management information system can be seen in many ways. They include better goal-setting, team-building,

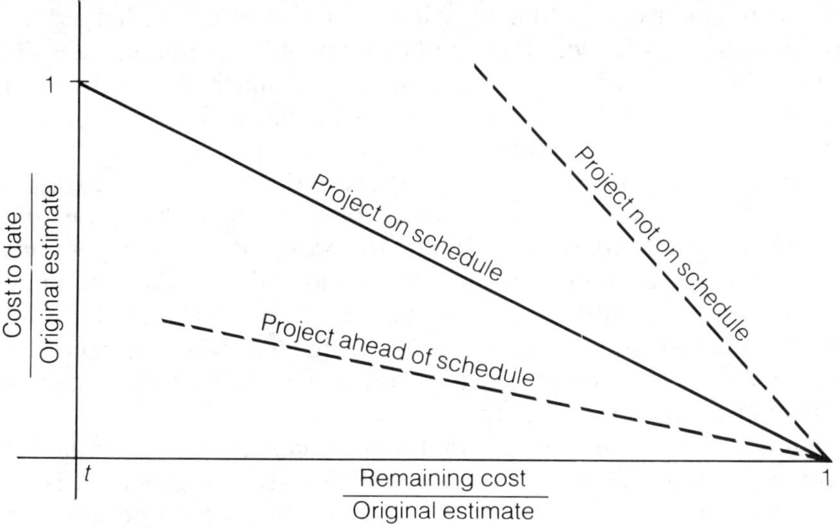


**Figure 5** Example cost-based progress chart.

commitment and resource allocation. They help to reduce the number of priority changes and to increase the honesty of reporting because the focus on milestone and on forward planning highlights the need for more detailed discussion when it is required. This provides individuals and teams with the autonomy they usually desire, i.e. they plan, organise and report on their own activities. At the same time it provides the support mechanisms so badly required when things are not going so well—sometimes owing to factors outside the project leader's and team's control. In this context it is important to remember the point that it may be necessary to leave some slack in the system to allow for as yet unanticipated activity. Research has consistently shown that performance is improved when scientists and technologists have clear goals, autonomy within agreed boundaries and adequate resources to carry out the work. The message is, clearly, don't overplan.

Finally, it must be emphasised that planning and monitoring should not be seen either as external to the activity being undertaken or as overheads. They are an essential part of management which, if done well, will contribute not only to the success of the project, but also to the performance development of the individuals involved.

## Performance Audit

At the end of the day two questions are inevitably asked: how effective is our R & D, and are we getting an adequate return? Research carried out by many people has always concluded that the answers to these questions are in the affirmative. However it is measured, and taking into account the failures as well as the successes, R & D has been shown to provide a higher return than that from most other parts of the business, and figures for public or social rates of return are even higher. The obvious point to be made is that this return is not guaranteed and is not usually obtained without considerable further effort. R & D provides opportunities that must be taken up and implemented for the returns to be achieved. And this usually requires a much higher level of expenditure, for example on production and marketing, and a good deal of commitment by management in other functional areas.

It is not surprising, therefore, that most of the early studies of innovation identified good communication, the presence of a product champion and an understanding of user needs as contributors to success. Practical experience also says that timing is important—too early may be as bad as too late, and real product or process superiority is an important factor to consider. It is, of course, easier to identify some of these in hindsight than when projects are being initially evaluated, but it is true that too often there is less 'learning from experience' than might be desirable.

It has been argued by many people that it is valuable, if not essential, to look back at the way in which projects and activities have been progressed—in other words, to carry out a 'post-mortem'—and that this should be done on successful as well as less successful projects, reviewing both the outcome and the process of management. If done well, this can provide very valuable information to help in the design and management of future projects. The methods and procedures outlined in the previous sections, the checklist, the planning and monitoring system, etc., provide very useful starting information for this purpose. For example, they allow early assumptions to be checked and actual progress to be compared with plans and initial estimates. They can be used to focus on costs, time, resource management, technical issues and, of course, on implementation and ultimately on benefits obtained.

It has been suggested that a useful but more subjective analysis of performance can be obtained by assessing individual projects on a scale that ranges from very poor to excellent. A five-point scale might include as intermediate values poor, marginally below expectations and equal to expectations. On this scale excellent would then correspond to above expectations. This approach can be used for ongoing projects, but it does not provide as detailed an analysis as a post-audit based on original assumptions, plans and estimates. An overall performance measure for a set of projects, a department or the laboratory as a whole can be obtained by multiplying the scale score for an individual project by a measure of its size, e.g. cost, and calculating a weighted average. Such a measure is most useful for monitoring trends over time.

An alternative and very useful procedure for assessing the performance of an R & D laboratory can be based on the relevance tree which was described in an earlier section. For this purpose it is necessary first to identify the objectives of the laboratory and then to identify the relationship of individual projects to these objectives. For example, it can be valuable to distinguish work of a long-term nature from that of short-term, to separate product from process development. Technical service and troubleshooting are likely to be other categories. A simple relevance tree might then look as shown in Figure 6.

There are many other ways of presenting this type of information, including, for example, a relevance matrix, as shown in Figure 7.

The body of the matrix contains some chosen measure, such as the amount of expenditure on the project or the number of man-hours planned. Total cost

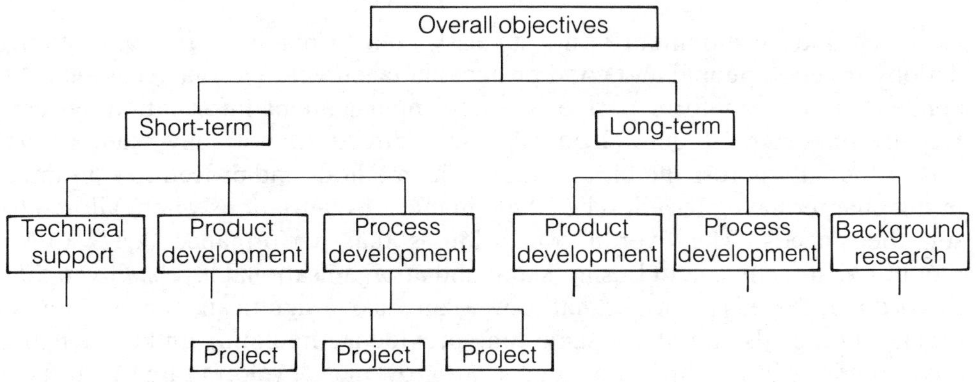


**Figure 6** Simple example relevance tree.

| Projects | Categories | | | | | |
|---|---|---|---|---|---|---|
| | Short-term | | | Long-term | | |
| | Technical support | Product development | Process development | Product development | Process development | Background research |
| 1 | | | | | | |
| 2 | | | | | | |
| 3 | | | | | | |
| 4 | | | | | | |
| 5 | | | | | | |
| 6 | | | | | | |
| etc. | | | | | | |

**Figure 7** Example relevance matrix.

or man-hours allocated to each category can be easily calculated, as can, from this, a view of the relative importance being given to the different categories. The actual expenditure over time and the progress achieved on individual projects can then be compared with each of the agreed categories and against the originally agreed allocation of resources to obtain a useful measure of overall performance.

## Conclusions

Research and development is not an easy area to manage. The basic foundations are good people and good project management. Various techniques, if applied wisely, can form part of a useful management information system. Equally important is good leadership and direction, not only from within R & D, but also across the organisation. The methods and procedures outlined in this chapter have proved to be a help in many organisations when tailored to suit their needs. The overall conclusion is that performance needs to be monitored at technical, at business unit and at organisational levels. Given that this is done, there is no doubt that R & D can make a significant contribution to organisation goals and at the same time provide a stimulating and challenging environment within which individuals can grow and develop in line with their personal needs.

## Further Reading

Allen, T. J. (1986) *Managing the Flow of Technology*. Cambridge, MA: MIT Press.

Burgelman, R. A. and Sayles, L. R. (1986) *Inside Corporate Innovation*. New York: Free Press.

Collier, D. W., Monz, J. and Conlin, J. (1984) How effective is technological innovation? *Research Management*, September/October.

Eres, B. K. and Raz, B. (1988) A methodology for reducing technological risk. *R & D Management*, **18**, no. 2, 159–67.

Gobeli, D. H. and Brown, D. J. (1987) Analysing innovation strategies. *Research Management*, July/August.

Gold, B. (1988) Charting a course to superior technology evaluation. *Sloan Management Review*, **19**, Fall.

Gunz, H. P. and Pearson, A. W. (1977) Matrix organisation in research and development. In *Matrix Management*, edited by K. W. Knight. London: Gower Press.

Krogh, L. C., Prager, J. H., Sorensen, D. P. and Tomlinson, J. D. (1988) How 3M evaluates its R & D programs. *Research and Technology Management*, November/December.

Pearson, A. W. (1987) Planning and control of research and development. In *Handbook of Financial Planning and Control*, 2nd edition, edited by M. A. Pocock and A. H. Taylor. London: Gower Press.

Pearson, A. W. and Davies, G. B. (1981) Leadership styles and planning and monitoring techniques in R & D. *R & D Management*, **11**, no. 3, 111–16.

Souder, W. E. (1980) *Management Decision Methods for Managers of Engineering and Research*. New York: Van Nostrand Reinhold.

Thamhain, H. J. and Wilemon, D. L. (1987) Building high performance engineering project teams. *IEEE Transactions on Engineering Management*, **34**, no. 3.

Tushman, M. L. and Moore, W. L. (eds) (1982) *Readings in the Management of Innovation*. London: Pitman.

Von Hippel, E. A. (1988) *The Sources of Innovation*. Oxford: Oxford University Press.

# 1.3 *The Management of Intellectual Property Rights*

*Keith Hodkinson*

## What are Intellectual Property Rights?

Intellectual property rights are essentially legal monopolies or quasi-monopolies designed to give protection to those who incur expense in R & D or who expend time and effort in intellectual creativity, so that they can recoup their costs and efforts and make profits by enjoying a legally protected market lead for a certain time. Thus authors enjoy protection through copyright from piracy of their works, inventors enjoy a market monopoly to exploit their invention, designers their designs. On a slightly different level, trade marks protect the reputation of the trader from imitation and counterfeiting of his goods.

Whether by enabling you to exclude others or by enabling you to license an invention or design to be used or produced by others, intellectual property rights are a valuable means of assisting in the exploitation and transfer of technology. This is particularly important to those without the capacity to satisfy market demand from their own production resources, such as universities and small companies which may license the invention to others that do have such resources. It is also important to those who cannot risk the investment in new machinery and plant if their new product is to be copied and undercut by those with better resources or cheaper labour, who did not invest the time and money in the original R & D.

The system also encourages disclosure of inventions through patent documents, which might otherwise be kept secret, e.g. through the use of black box and secret process technologies, and thus assists incremental development in many fields. This in turn makes available through the processes of licensing-in of technology the means for technically backward companies to catch up or to improve their own product range.

A variety of intellectual property rights exist to protect different aspects of research and development and these will be briefly considered.

## Patents

### *What is a patent?*

A patent provides the legal right to exclude others from exploiting an invention that is a technical or scientific development having some practical applicability and which is described in the patent. In most countries this right, once granted, will last for up to 20 years from the date the patent was applied for, although in the USA and Canada the period is 17 years from the date of grant of the patent (which may be three or four years after the initial application).

Patents are national rights, so that a patent granted in the United Kingdom does not give the patentee any rights in, for example, the USA. Therefore, for international protection separate patent applications must generally be filed in all the countries in which protection is sought. There are now a number of ways of cutting down on the expense and effort involved in such a campaign, as a result of the European Patent Convention, which provides for a single European patent in 13 countries, and through the Patent Co-operation Treaty, which enables co-ordination of patent applications in 42 countries through a single examining body.

### *What are patents granted for?*

Patents are granted in respect of inventions that are: novel, in the sense of not being previously known to the public; inventive, in the sense of not being obvious developments over existing technology; and capable of industrial application, i.e. not merely scientific discoveries, mathematical formulae or means of doing business or performing calculations.

Most novel and inventive processes, methods, devices and products will be patentable. However, most countries have certain public policy exclusions to deprive certain inventions of protection. Computer programs are commonly excluded from patent protection, although if the program has a technical or industrial application, such as controlling the operation of a machine robot, protection will be available. Another common exclusion is that of methods of treatment of the human or animal body, although a drug used in such a method will be patentable. Inevitably, national patent laws differ but the position in most countries in Europe is set out in Table 1.

The important point to make is that the layperson should never make a judgement as to whether a particular invention is or is not patentable. This is a question on which professional advice of a patent agent should be sought. The patent agent can conduct searches to determine whether the idea has been published before and whether there is any difference from existing knowledge that would give rise to a patentable claim.

**Table 1** European patent requirements and exclusions

**Fundamental requirements for patentability of invention in Europe**

1. Novelty—invention must not be publicly known anywhere in the world
2. Inventiveness—invention must be non-obvious to person skilled in the field
3. Capable of industrial (including agricultural) application—invention must not be a mere theorem or abstract concept

**Express exclusions from patentability in Europe**

Discovery, scientific theory or mathematical method

Literary, dramatic, musical or artistic work or aesthetic creation

Scheme, rule or method for:

- performing a mental act
- playing a game
- doing business

Computer programs (contrast an industrial machine driven by one)

The mere presentation of information

Offensive, immoral or anti-social inventions

Varieties of animal or plant or any essentially biological process for the production of animals or plants not being a microbiological process or product of such a process

Methods of treatment of the human or animal body by surgery, therapy or diagnosis (contrast product for use in such activity)

## *Should we bother with a patent?*

What is said above does not mean to say that a patent agent should be consulted over every invention. This would involve substantial unnecessary expense and waste the time of the invention and the patent agent. Many inventions simply do not merit the expense of a patent in commercial terms, or will enjoy better alternative forms of protection. Equally, to ignore patents as a matter of policy is folly. A middle-ground strategy can be set out in the form of a series of questions that will lead to rational decisions being taken in the individual case.

### Is what we have thought to be new?

A very substantial number of inventions are not new: they have been invented before by someone else somewhere else. Patent agents can do fairly sophisticated prior art searches to help determine the answer to this question but much preliminary checking can be done by the inventor him or herself.

### Is what we have clever?

If it is, it may well be new and inventive and may well be potentially commercially attractive, e.g. if it presents an easy way of doing something. The fact that it is simple is not necessarily material. Rubik's cube was simple but it was patentable and clever. So was Ron Hickman's famous 'Workmate' workbench. Both Rubik and Hickman became multimillionaires on the basis of their patents.

### Is what we have commercially attractive?

There is no point in having a monopoly over something nobody wants. The object of a patent is to assist in the exploitation of an invention. To have a patent does not guarantee commercial success—indeed the majority of patents are commercial flops, much to the chagrin of the private inventor patentee, who laments that the world does not beat a path to his door and ignores his genius.

### Is it difficult to keep it secret easily when exploiting it?

Some inventions, such as a secret formula, may be easily kept secret. In such cases there may well be an argument for secrecy, especially if enforcement of the patent would be difficult. But serious thought has to be given to how easily reverse engineering analysis or simple industrial espionage could be utilised to uncover the secret.

### Does it have a lifespan of more than, say, five years?

If not, then the cost of the patent may not justify the short term of protection needed. On the other hand, high profits and short lead time in the industry in question may demand this type of protection.

If the answer to all five of the questions set out above is 'yes' then consideration should be given to seeking patent protection in at least the most important markets for the product or process concerned. However, a word of warning is justified. Do not over-estimate what a patent can do. It does not in itself ensure success. It merely gives a head start for rather a long time. It gives a monopoly that others cannot break. Even if they come up with something better they may need consent to produce it because it incorporates a patented invention. So the patent buys a market lead in production and sales that far exceeds that which merely being first in the market-place would provide. But most inventions are commercial failures. Patents are an assistance and not a guarantee of success.

For someone with a viable invention and no production resources or too little money, the patent can give options that would not otherwise be available. A patent makes it easier to sell the invention, to license the invention or to put something of value other than cash into a joint venture. And it helps to keep the competition at bay for a time. In exceptional cases the competition may even have to buy off the patentee, or buy him or her out. All of this presupposes that the invention is worth having in the first place.

Obviously the decision to seek or to maintain a patent is a commercial one. Look at what is going to come out of it and what it is going to cost and make a decision. You can always pull the plug on the expenditure at any time and you should not be afraid to do so. But you should also remember that once you have pulled the plug you cannot opt in again—if the invention that was initially unsuccessful becomes a winner you cannot reapply—you have lost the monopoly for ever.

### *How do we get a patent?*

Three crucial points must be made. The first is that in almost every country any non-confidential disclosure of the invention to anyone before the date of filing an application to patent the invention will fatally prejudice the application. The golden rule is: file first and disclose after. The USA and Canada have some anomalous exceptions but these cannot be easily relied on by foreign inventors and only apply within those countries.

The second point is that in almost every country the first to file and not the first to invent will have the better right to a patent in most cases: if two inventors come up with the same idea independently of each other the first to file will have the better right to the invention irrespective of who actually was the first to make the invention (again the USA and Canada are anomalous exceptions, but Canada is now coming into line with the rest of the world on this point).

The third point is that if it is subsequently decided to keep the invention secret, a patent can be withdrawn at any time up to 18 months after filing (in most cases) without it having been published, so that secrecy can be maintained. Mere filing of an application is not fatal if a decision is later taken to withhold the invention from public disclosure. After 18 months the invention will be published for all to read.

In fact published patents can be a fascinating source of information. In recent years there was much press speculation over such matters as the 1986 America's Cup winner's racing yacht keel and the pilot's helmet stolen or photographed at an airshow in the UK in 1988, but these so-called secret designs were available for inspection in patent documents in the UK Science Reference Library.

In most countries a patent is obtained only after rigorous examination of the application for novelty and inventiveness, and compliance with legal formalities by the Patent Office. The application contains a summary of the existing state of the art, a description of the invention and how it is used and claims to legal monopoly in those aspects of the invention that are new and inventive.

In some countries, such as the USA and Canada, the inventor must disclose the best method known to him or her of putting the invention into effect, whether alternative methods are available, and any prior inventions of which he or she has knowledge or ought to have had knowledge which affect the consideration of whether the invention is novel and inventive.

The patent application is a complicated legal document as well as an exposition in scientific or technical terms of the invention. Unless the applicant is well versed in patent practice it is extremely unwise to attempt to draft and file the application without professional assistance if the invention is likely to have any commercial significance. A badly drafted application can lead to an otherwise patentable invention being rejected or, even worse, held invalid in subsequent court action; it can lead to loopholes in the scope of claims and thus narrower protection than the applicant would otherwise be entitled to. An application that is too narrow cannot at any later stage—even before grant—be expanded to cover things not previously disclosed. Conversely it may disclose things that need not have been disclosed and that would be better kept secret.

Finally, in some countries, such as the USA, a failure to disclose all that should have been disclosed can lead to loss of the patent and even to liability for fraud by the US Patent Office. If overseas protection is sought the assistance of agents in obtaining adequate translations into the local language and in dealing with patent office objections will be in practical terms essential and in many countries legally required.

### *How much does it cost to get a patent?*

A typical invention for a mechanical device might cost in the region of £1000 to £1500 to patent in the UK, and there are annual renewal fees to pay after the fourth year, which rise to as much as £200 per year in the final years of the patent. For coverage in, say, most of Western Europe the cost would typically rise to £5000; if the USA and Japan were added the cost might rise further to, say, £15 000. In fields such as biotechnology, where professional expertise is highly specialised and more expensive, the costs could be higher. A typical time/cost graph for typical country combinations is set out in Figure 1.

The applicant is not committed to all these costs merely by applying, let alone all at the time of filing. A so-called 'provisional' application (a term with no technical meaning under the new law, but a hangover from the old legislation and one still frequently employed) may be filed in the United Kingdom, and will give the applicant a period of 12 months to decide which countries if any he or she wishes to proceed in when seeking a patent. The cost of this for the typical mechanical invention referred to above might be in the region of £350 to £400.

Even at the 12-months stage many of the eventual costs of an international campaign can be substantially deferred to, say, 30 months after filing in the United Kingdom, by which time the true commercial worth of the invention will be better known. At any stage it is possible to terminate expenditure and abandon some or all of the countries originally selected.

Even if the invention is not patentable there will often be alternative or supplementary forms of protection available and these often cost less.

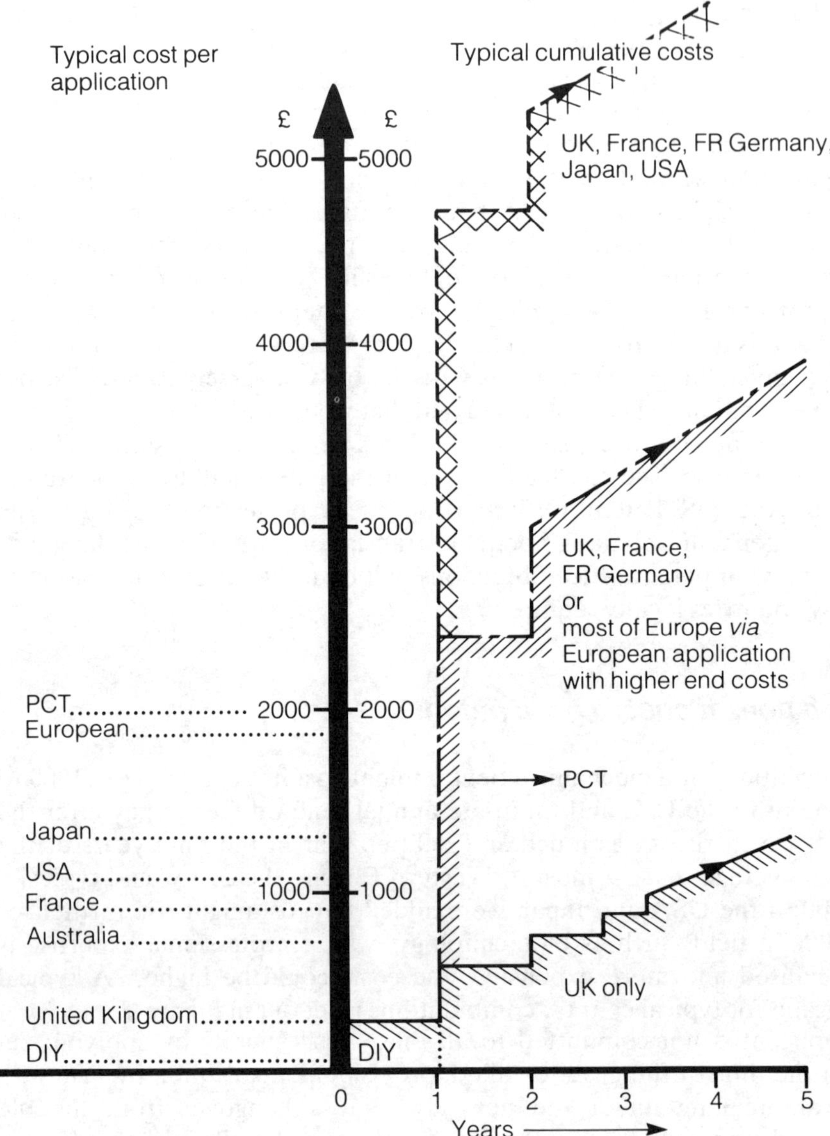


**Figure 1** The costs of patenting (PCT is the Patent Co-operation Treaty).

## Design Rights

### *What is a design?*

For our purposes designs may be regarded as aspects of shape, configuration, pattern and ornament to industrially mass-produced goods. In some countries both functional (non-patentable) and purely aesthetically appealing designs may be protected either by registration or even by reliance on copyright or allied rights in design drawings and prototypes. In others, only non-functional designs may be protected in the absence of a patentable functional feature. There is very little international harmonisation of the law on designs, unlike that for patents, and professional advice is essential if international protection is desirable.

As will be obvious, designs, like patents, are national rights, so that a design registration or unregistered design right, granted in the United Kingdom do not necessarily give the designer any rights in, for example, the USA. However, there are no means yet available for cutting down on the expense and effort involved in obtaining international rights. However, the cost of international design protection is generally somewhat less than that for a patent and a typical cost of £250 in the United Kingdom and £300 to £500 in overseas countries might be anticipated.

In the United Kingdom the law of individual designs—which had been unsatisfactory for many years—has recently been reformed. It is too early to tell how the new provisions will affect industrial design protection in practice but the basics of the new regime can be explained, in simplified form.

Most aesthetically pleasing industrial designs will enjoy a choice of two forms of protection: many designs can be registered if they are novel—meaning not publicly known in the United Kingdom before the date of the application—which will give a period of 25 years' protection from imitations; or it will be possible to rely on copyright in the original drawings for a two-dimensional design for a period of 25 years even without registration, which is a cheaper (i.e. free) form of protection but one that will be weaker in some regards. Some designs incorporating features of shape, even if aesthetically pleasing, will only enjoy the same degree of protection as functional designs, however.

For functional designs, such as a new configuration of exhaust pipe, the period of protection will be 15 years from the creation of the original design document or prototype, or 10 years from first marketing of the product (whichever is the shorter), under the new (free) unregistered design right. After five years other people will be able to reproduce the design if they offer to take a 'licence of right' for a reasonable royalty payment. Thus a patent on a patentable functional item will be very much more valuable.

These new regimes are being phased in gradually over the next decade. Professional advice should be sought in all cases.

## Plant Seeds and Varieties

Some countries protect plant seed and varieties by a patent but others, such as the United Kingdom, maintain a separate regime to protect the reproductive material of the new variety from commercial reproduction. Methods of cloning and other biotechnological methods will usually be protectable by patent. Recently in the United States genetic methods associated with genetic manipulation in mouse breeding were patented and similar methods would be *prima facie* patentable in respect of plant life.

## Trade Secrets

These are not strictly intellectual property rights at all but remain an option, particularly for short-term protection, such as when deciding whether to patent or not. Most countries give some degree of protection against abuse of trade secrets but do not protect the possessor from accidental disclosure or from the independent discoverer in the way that patent and designs do.

There may be a number of reasons for looking to trade secrets. The invention may not be patentable but may be useful commercially and easily concealed, such as in a black box or factory secure area. On the other hand, the vulnerability of the information to leakage may lead the owner to conclude that the longer-term security of a patent may be necessary. Equally, the licensing of trade secrets and know-how can be extremely difficult to accomplish satisfactorily.

## Copyright

Copyright remains in most countries the principal form of protection for material such as computer software, although increasingly computer firmware is being accorded special protection through autonomous regimes, such as the Semiconductor Chip Protection Act in the USA and similar regulations in the UK and elsewhere in the European Community.

In most countries copyright arises automatically and without formality as soon as the work in question is created but in the USA and most South American countries a deposit of at least extracts from the program are required for full protection in litigation. Expert advice should be taken in all cases.

Most countries do not require any express claim to copyright, although the required claim in UCC (Universal Copyright Convention) countries of the copyright symbol (©) plus the name of the owner and year of publication is common practice throughout the world. It is certainly prudent to make such a claim on the program. The USA became a party to the Berne Convention on copyright protection from 1 March 1989, which has led to reforms in the US law in this area.

Some countries will give a degree of protection to industrial designs through the protection of copyright in the production drawings or an original concept drawing of such designs. Expert advice should always be taken on any particular case and the law is haphazard at an international level.

## Non-Technical Intellectual Property Rights—Trade Marks

A trade mark can often be a useful commercial supplement to a patent or design. It will be associated with the product or process. Even when the product is not new, not patentable, not anything, it can still be a market leader because it is better built, better designed, better everything. The name then sells the product. The trade mark symbolises the reputation for quality or reliability of the product or process and long after the other rights are out of force it may be a valuable right worth substantial sums. Thus a well-known trade mark helps to preserve product differentiation even in a crowded competitive market and assists in the development of brand loyalty and in dealing with counterfeiters and other competitors seeking to cash in on a market developed and led by the innovator. The roles of trade marks should not be forgotten when planning commercial development of technology, even through licensing.

Trade marks should be registered whenever possible. Although in many countries there is no need to register a trade mark and rights can be acquired by use and reputation in the mark a registration is nearly always stronger and cheaper to enforce. Registration of your trade marks is cheap and can operate as a very effective insurance policy against legal costs, because it enables you to cut the cost of litigation founded on your reputation and, in certain cases, to invoke the assistance of the public authorities, such as the Customs and Excise, in cutting off counterfeit goods at source.

Furthermore, the risk of accidental trade mark conflicts arising is reduced because registered trade marks are easier to search for than unregistered marks and a reputable company, especially a reputable foreign company selling into the country for the first time, will search for availability of the proposed mark before adopting it. Equally, of course, it is always wise to have a trade mark availability search conducted yourself before adopting a new trade mark.

Trade mark registration is a national affair and the period taken for a mark to be registered varies from as little as a few months to three or four years, depending on the country involved. Once the mark is applied for there is no reason to refrain from using it and in some countries early use before registration is extremely desirable. Your patent agent or trade mark agent will advise you of the requirements of particular countries.

## Management Issues After the Decision to Seek a Patent

Obviously the granted intellectual property right requires management in a number of ways. However, many important management issues arise at a much earlier stage.

The first obvious issue has already been raised and considered: where a new invention comes to light do you want the protection of an intellectual property right and what form of protection is most suitable? Once this decision has been taken cost budgets and the countries in which protection is to be sought must be decided upon, with regard to the commercial practicality.

A second issue that has to be considered, and one that is often neglected, is whether the proposed applicants are actually entitled to the right. This is rarely a problem where the invention is an internal project conducted by employees of the company or university in question. The employer will generally own the right if the invention was created in the course of employment by the employee. Where the project has been funded by a joint venture or government grants there may be questions of entitlement to be resolved, and this is also true of developments made by staff who are not employed to invent or who have done something peripheral to their activities, or by research students. Equally, in most countries there is no such thing as shop rights in inventions such as exist in the USA, although the myth of the shop right lives on in many companies. Reliance should not be placed on the terms of an employee's contract of employment in such cases because the law sets out stringent restrictions on the circumstances in which employees can be deprived of inventions they would otherwise own and on acceptable pre- and post-invention contract terms in this field.

Third, internal administrative organisation can considerably enhance the prospects of success and reduce the costs of intellectual property acquisition. The inventor team will be aware of relevant technology that might affect the patentability of the invention; their notebooks and other explanations will assist in the prosecution of the applications and in dealing with objections, and will also determine who is truly the inventor or inventors.

Most countries require accurate statements of inventorship for patents at least and a failure to name correctly the inventors may prejudice the enjoyment of the right acquired. This is critical in, for example, the USA. For this reason egos have to be replaced by truth and if a junior research fellow made or contributed to the invention or a senior professor did not do so this must be reflected accurately in the inventors' designation, especially in research team inventions.

Fourth, much potentially valuable material is lost through over-enthusiastic publication or disclosure or simply through the significance of a particular activity being missed. Internal management should include regular review of research activity and encouragement of staff to report potentially useful research results well before publication to ensure that proper consideration can be given to the issue of intellectual property protection.

Industry has long taken this approach but universities have lagged behind. This is largely a question of simple education, although some academic staff will always consider the glory of publication more important than the money—until the significant loss becomes apparent. Some very successful academic institutions, such as the Massachusetts Institute of Technology, have created model regimes which are being adopted throughout the world in dealing with purely internal and outside-sponsored research.

## Obtaining a Patent—the Prosecution of a Patent Application

As soon as your application is on file you can disclose it without prejudicing your rights. You can tell the world and say there is a patent application pending. But you might not want to do so. Until shortly before publication of the patent application, 18 months later, it is still secret and you can prevent publication by withdrawing the application before that time if you so wish. This might be because you have discovered it is not inventive or because you want to rely on trade secrets instead.

Ultimately the price for getting a 20-year monopoly over your invention is that you have to tell everyone how it works. The invention must be described adequately in a document that is eventually published by the patent office and available for all to read. Failure to do so can lead to invalidity of the patent, and in the USA you and your agent could in theory suffer prosecution for fraud by the US Patent Office if you do not disclose all you know of relevance, including what you know other people have done. However, in the UK and Europe at least, you do not necessarily have to give away all you know about the best method of using the invention, as you once did and still do in the USA. The associated know-how need not be disclosed. In practical terms, of course, your patent agent needs to know as much as possible to do a good job for you.

When considering whether to grant you a patent, the patent office sets strict deadlines within which responses to their queries and objections must be given and demands for government fees must be met. If you miss a deadline you may well lose the patent. Too many private inventors lose applications through this and too many people employing patent agents cause themselves extra expense by forcing the agent to chase them for information on time. Even if you are in a hurry, and even though you have to meet their deadlines, the patent office will not necessarily meet yours! It can take anything up to four and a half years before the patent is finally granted, although two and a half years is more usual in the United Kingdom and Europe. Nothing prevents you from going into production before then, but you must make sure that you applied for the patent before starting to make and sell.

## Managing Granted Intellectual Property Rights

Once intellectual property rights have been granted a number of tasks face the owner. These tasks include: defence and active exploitation of the rights; monitoring of other people's activity; and the more mundane administration and finance of the rights themselves.

Taking the last item first, from the fourth year of its life onwards there is an annual renewal fee on a patent in most countries. If this is not paid on time there is in most countries a penalty fee payable in addition, and after a delay of, usually, six months the patent can be lost for good in the absence of highly exceptional circumstances. Most patent agents operate reminder services of varying efficiency. There is a particular problem overseas because of delays that can result from currency exchange and related problems. Far too many patents lapse accidentally. Similarly, renewal fees are payable on registered design and trade mark rights and similar principles apply. Budgets have to be set to take account of these renewal fees.

At the stage of grant some consideration may be given to patent and similar watching services, which will monitor applications by third parties in which your patent is cited or that are otherwise relevant to the technical field covered by your patent or other right. These can be useful in keeping you up to date with further developments or with possible infringments of your rights, new avenues for exploitation of the right and alternatives to your methods or products.

## Exploitation of the Rights Obtained

Exploitation of intellectual property rights is a convenient way of profiting from technology because it provides an easily identified and defined package of information and know-how to transfer to value and to control. The legal mechanisms available for exploitation fall into three basic forms: exploitation by self-production or use of the technology; sale and assignment of the intellectual property rights; and licenses granted to third parties. These should be considered in turn.

If self-production is the most likely option the use of intellectual property rights will be principally defensive, to keep other traders out of the market place so as to preserve the benefit of a monopoly over production and a market lead over some desirable piece of technology. In such cases most attention will be given to the monitoring of the market place to ensure that third parties using the technology are detected and pursued. This issue is considered below.

There are alleged instances, some more myth than fact, of companies seeking to use the patent to keep a new piece of technology off the market place altogether, on the basis that it threatens their own product base or they cannot exploit it as they would wish. In such cases and in cases where the owner refuses to grant licences or otherwise adequately exploit the invention except on unreasonable terms, most countries provide for orders of compulsory licences to be granted on terms decided upon by the courts.

Frequently it will be impossible to match demand or to exploit to the fullest the technology in question through self-production. The world is a big place. Accordingly, exploitation through third parties may be necessary or desirable. As mentioned briefly above, in some cases it will be imposed by the courts on the request of a third party or the government. In yet other cases the national interest may dictate that a particular invention be exploited for the defence of the country and compulsory licences may be granted on reasonable royalty payment terms.

Such exploitation through others, whether voluntary or involuntary, will generally be controlled through licensing of the intellectual property right and the terms and conditions on which exploitation is permitted will be regulated by a written agreement or order. A licence will provide a consent to use the invention or design in return for a royalty payment or other form of compensation. There are a number of major advantages to voluntary licensing:

- turnover can be expanded more rapidly to meet demand
- profits can be made without personal investment and risk
- local licensees can monitor/service the market more easily

- local licensees can adapt to local circumstances
- import tariffs and similar levies can be overcome.

Wherever voluntary licensing is considered a general point applies and this must be stressed. This is no more the occasion for a do-it-yourself approach than is the preparation of a home-made will (or a patent). Expert advice must be sought from a patent agent or from a solicitor specialising in these matters when seeking to draft agreements. Three factors demand this, and are given below.

First, there are legal restrictions on the terms of agreement that can be reached, breach of which, particularly in the USA and in the EEC, can lead to an inability to enforce the rights against the licensees or even third parties. Some very attractive commercial terms, such as tying clauses, under which a condition of granting the patent licence is that the licensee buys non-patented items from the licensor, and clauses requiring the licensee to assign improvements back to the licensor, are prohibited. Secondly, the structure of the agreement can be complex and will in any case of dispute be determined by a court of law. A badly drafted agreement can leave loopholes and cost money. Specialist advisers will have become experienced in the problems likely to arise and can cater for them in the agreement. Thirdly, tax implications loom large here, as in all commercial agreements. Specialist advice is needed, particularly in international agreements.

Because of the very complexity of these issues this chapter deliberately avoids discussion of the substantive issues to be faced. Professional advice must be sought.

As an alternative to licences under the intellectual property rights, some people, especially private inventors who might find the involvement in an ongoing licence too much, simply assign the right in the same way as assigning a debt or other piece of intangible property. This might be for a single lump sum, for a royalty on turnover in the right assigned (just like a licence) or for some combination of the two. In practical terms there may often be no difference between an assignment and an exclusive licence that concedes the right to use the patent or other right, even to the exclusion of the owner, and the law frequently treats the two as equivalent. Technical and legal factors, as well as commercial factors, will determine the choice.

## Defence of Intellectual Property Rights

The defence of intellectual property rights is just as important as exploitation through licences and the like. It may involve litigation, although the vast majority of intellectual property disputes are settled by the parties outside court, either through the defendant refraining from infringement or through licence agreements being reached.

To stop someone from infringing the right often requires a threat of litigation although nine out of ten cases are settled around a table before getting to court. You have to take the initiative—in most countries the patent office will not help you and nor will the police. It is in most cases a purely private matter. Professional advice must be taken before making such a threat because provisions exist in many countries enabling a person who is the victim of an unjustified threat to seek damages against the person making the threat.

If yours is not one of the nine out of ten cases that settle, litigation and the subsequent court cases can cost a lot but it is increasingly possible to take action before the patent office itself, rather than the court, in a wide range of conflicts. Like all litigation, the protection of intellectual property rights through legal actions in court against those who infringe the rights, or who challenge the validity of the rights, can be expensive and the work involved in conducting the case can be substantial: the professional advice does not come cheaply, although figures vary dramatically throughout the world and are often much exaggerated.

In most cases it is now possible to insure against intellectual property litigation costs, either through a general litigation policy or through specialist insurance policies devoted to intellectual property matters. As a typical example, a UK cover level of £100 000 (more than enough to cover most cases) may be had for, say, £150 per annum, as long as certain preliminary checks on the validity of the right to be insured are conducted first—the equivalent of a medical for life insurance. Because of the terms and the premium structures of the policies involved, cover should be sought as early as possible in the life of a patent application or similar right and should certainly be reviewed when looking at product liability cover before launch.

## Living with Other People's Intellectual Property Rights

Even if you decide that patents and the other intellectual property rights are not rights of which you wish to take advantage you cannot ignore them. Other people's patents and other rights can cost you money. Even other people's invalid patents can cost you money if they sue you, because you have to defend yourself and establish the invalidity of the patent or other right on which you are sued. To enter into production with a product that claims to be subject to someone else's patents is sheer folly unless very careful thought has been given to the matter.

Equally, all may not be as it seems. Many people are daunted by the apparent threat of a claim to be infringing a patent and do not take the trouble to investigate the matter fully. How many supposed patents are in fact expired or lapsed through non-payment of renewal fees? How many are simply invalid or do not cover what the owners claim they cover? How many patent pending cases are pending overseas and not in the UK? If you think a patent may be in the air, investigate. It does not cost a lot to do so and a little investigation can save a lot of expense in the longer run. Similar advice is applicable to the other intellectual property rights.

Do not be afraid of patents and the other intellectual property rights. It was once said of deeds that they were difficult to read, impossible to understand

and disgusting to handle. Once that may have been true of patents too, but no longer. They have an aura of mystery that is quite unjustified. They are, when all is said and done, technical documents with a legal element added on to the end. The bulk of a patent is a concise and clear account of an invention, with illustrative drawings complying with the usual engineering drawing standards.

Such a patent and other intellectual property right documents can contain a great deal of useful and new information about technology and about competitors. You might need this information for a number of reasons. You might want to know what your competitors are up to. You might want to start to catch up with the competition and think of an improvement. You might want to look for ways of avoiding the patent. You might want to take the risk of copying and infringe. You might want to look for ways of attacking the patent.

## Summary

I have already said that even if patents are not thought to be appropriate for your institution or company, you ignore them at your peril. If you are not interested in a monopoly and just want to prevent other people from being able to stop you from doing something or to threaten you with such action the answer is cheap and simple. Just disclose the invention to the public. The same applies for registered designs. Unregistered design rights need not be disclosed to safeguard your position but of course much that is protectable by such rights will also be potentially patentable so that non-disclosure may be a risk.

If you want to be sure that your new product does not infringe third party rights then it is possible to do some very rapid and cheap searching of the patent literature to determine whether you infringe existing patent rights. It is also possible to undertake clearance searches for trade marks, but it is more difficult to check-up on registered designs and almost impossible to run any check on design copyright.

If you do face a problem you can always get advice on avoidance engineering—for instance, you might have a prototype that is a bit close to some else's machine. You can obtain professional advice on changes to the design and on modifications to avoid other people's design rights. Again the important thing is to make sure you know what the true position is. Advice is relatively cheap and quick. Ignorance may in the short term be bliss but in the mid- to long term it is often much more expensive.

## References

Hodkinson, K. (1987) *Protecting and Exploiting New Technology and Design*. London: Spon.

Hodkinson, K. (1986) *Employee Inventions and Designs*. Harlow: Longman.

# 1.4 *Managing Scientists and Technologists*

*Keith Alsop*

## Introduction

Research is still—and always will be—labour intensive. Around two-thirds of the cost of research is attributable to wages and salaries. Development is somewhat less labour intensive, but the human input is still costly. In both research and development, it clearly follows that human resources are crucial. The way we recruit, deploy and train them, and the way we give them opportunity and encouragement to develop are important factors in the successful management of technology. These are the concerns of this chapter.

I start by considering the recruitment of scientists and technologists into an R & D organisation, and their induction into the working environment. I go on to look at the development of their careers and their motivation. Then I discuss the formation of successful teams, linking this with a useful model of the research process and drawing on recent research. Finally, I consider the maturity and staleness of teams.[1]

## The Individual

### *Selection*

If the thesis in the introduction—that in R & D the human resources are crucial—is right, then we need to pay great attention to recruitment, to ensure that we take on the right people. In passing, it is worth remarking that in the current (1989) situation, shortages in many disciplines mean that organisations are having to make themselves as attractive as they can to potential recruits rather than having the luxury of selecting who they want from a wide range of attractive and gifted applicants. Thus it will be important to give applicants the opportunity of seeing the laboratory in which they will be working, of talking with near-contemporaries in the laboratory, of discussing recent achievements of the laboratory and the use made of them, of seeing the library, and so on. Full information about the research and development operation and about its parent organisation must be sent well ahead of any interviews, so that the interview can concentrate on the applicant's work behaviour, scientific attainment, personal characteristics, etc.

The most successful selection process is always concerned with filling specific vacancies, and I shall consider this first, before looking at more generalised recruiting activities. The starting point is obviously the vacancy itself.

Assuming that a vacancy does exist, that the work still exists and that it cannot be covered through reorganisation and must be filled, the first step is to produce a job description, bearing in mind that this is a snapshot at a point in time and will, subsequently, change. (Hence the need, during a selection interview, to explore candidates' flexibility and potential.) The job description will then be turned into a person specification: i.e. what personal characteristics, knowledge, working habits, potential and experience must a successful candidate possess? Briefly, the job description and person specification will form the basis of the advertisement, inviting people to apply for fuller information and an application form. From the application forms, a short list of applicants to be interviewed will be established, and those applicants will be invited to an interview (Fletcher, 1986).

In preparation the interviewer must become fully familiar with the information on the application form, and identify from it areas that need further exploration. He or she should review the person specification and extract from it those characteristics that are essential to the performance of the job, and those that are merely desirable, in order to minimise the risk of over-valuing simply desirable characteristics and failing to notice the absence of some essential ones. This will also help to avoid the recruitment of over-qualified people, resulting either in an unbalanced workforce or in a high staff turnover.

Many person specifications in the R & D field include such characteristics as 'flexibility', 'drive' or 'creativity', which tend to be ill-defined and global. In order to interview effectively, these terms must be closely defined. In what circumstances is creativity, drive or flexibility needed? What sort are we looking for? In what circumstances is it important? Do we value it at all costs? Why do we require it? Can we set limits? And so on. The more we can be precise in our specification, the more readily we can devise a strategy to seek evidence on which to base our evaluation. We uncover the evidence by exploring the candidate's work history and by eliciting actual examples that will throw light on the extent to which the qualities we need have been exhibited in work situations. We begin with general open questions, the responses to which will invariably give several leads, each of which we can pick up in turn, following each to the depth and degree of specificity we need in order to decide. Always, we seek actual examples that illustrate what the candidate is saying, and always we seek evidence of *how* the candidate approaches work, as well as what he or she has done. Always our questions are purposeful and short.

The extent to which we are able to elicit relevant information will depend on building rapport—the situation in which sufficient confidence is established in a candidate for him or her to feel able to talk freely without incurring the judgement of the interviewer. The interviewer therefore must remain neutral, being warm towards the candidate without becoming over-friendly.

Panel interviews are different from the one-to-one situation. It is difficult to establish rapport between a candidate and one interviewer and more difficult between a candidate and a panel. In addition, each panel member normally has insufficient time to question at the depth necessary to obtain the required information, so consecutive one-to-one interviews are often preferable. Both, however, require that the interviewers agree between themselves how to divide the total field.

It is essential to train oneself to suspend judgement on candidates until the interview is complete. There is much evidence that if initial prejudice is formed, subsequent evidence is interpreted to support that prejudice and, indeed, that the course of the interview will be steered unconsciously to support it.

Once the interview is over, and not until then, we evaluate the candidate against the essential criteria we established at the outset. We need to be on our guard against being over-influenced by desirable—but not essential—qualities.

Finally, the most frequent reason for graduates leaving jobs after a year or two is that what they took to be promises made when they were recruited (or general statements about prospects, conditions of work, etc.) were unfulfilled. We need to be very careful to be realistic in what we say. The costs of recruiting staff who stay only a couple of years are very high; so are the costs of recruiting the wrong staff.

## *Induction*

Adequate arrangements must be made for receiving the recruit on the first day in the job. These must be followed by two important steps. The first is to make sure that there is a worthwhile task to be started, and that there is adequate guidance available so that an actual start can be made. Many recruits are greeted by the formula: 'We want you to work in such and such a field. Here's a list of references. Read them—and any others you may come across—and we'll have a chat in a couple of weeks.' It will always be necessary for a recruit to read him or herself into a new field of work, but alongside this he or she needs to have a task to do and should be given a role in an existing team so as to contribute to its progress. In this way the recruit will begin to be familiar with the working practices and culture of the organisation. In parallel, arrangements must be made to provide formal information about how the organisation works; how the recruit fits into its structure; what authority he or she has, etc. General and specific working policies must be explained. It may be helpful to go over the job description with the recruit.

This assumes that new recruits will be part of an existing team. Generally, this should be the case. Even if someone has been recruited to start some entirely new work, that person should start at least part-time in an existing and effective team so as to have the benefit of support in the first few weeks.

## *Career development*

The ways in which careers typically develop in an organisation may well influence the criteria used in the selection process. We may need to look for other qualities in addition to those required for the specific post for which we are recruiting.

Careers can develop in the following ways:

1. We may be able to offer a lifelong career in R & D. Not many R & D organisations can do this for all their staff, or even a majority. Even here there are two possible paths, one remaining essentially as an R & D worker, perhaps in a highly specialised field of work, and the other moving up in the managerial levels in the R & D organisation.
2. After a period in R & D, staff may move elsewhere in the organisation—into technical service, marketing or production. Some organisations quite deliberately use the R & D function as the main agency for recruiting virtually all their graduate intake, relying on subsequent transfers to fill posts elsewhere.
3. An organisation may recognise that some R & D staff will, after a few years, decide to further their careers by leaving, because neither of the routes for development just outlined seems available to them. While this does provide the opportunity for recruiting fresh graduates, it has little to recommend it overall.

It is important for an organisation to establish which of the development routes it will be using, or, more realistically, the rough proportion of staff that are likely to be able to develop along 1 or 2, and to begin assessing the potential of individual R & D workers for these types of development. It is necessary, therefore, to look for clues from the way in which research staff approach their tasks (both technical and managerial), the extent to which they handle complexity, and the time scales with which they seem comfortable. We must avoid forming a view too early and, as a consequence, over-accelerating or failing to provide opportunities for development.

While scientists and engineers working in R & D should be managed by scientists or engineers, the higher in the management hierarchy a scientist or engineer climbs, the less he or she will know about the scientific or engineering content of the work of his or her subordinates. Even in the manager's own field of specialist interests, new graduates are likely to bring the new knowledge. Increasingly, the manager's contribution will be concerned with broader aspects of the work of the laboratory, with judging the practicability of research proposals and the likelihood of success in the light of the estimated resources. This inevitably hinges on the view the manager takes of his subordinates and of their capabilities. But it does also suggest that we may need to take steps to broaden the experience of many of our staff, with the possible exception of those for whom we predict an increasingly specialist role. I will return to the question of regular moves in the later discussion of teams. If, however, we recognise the possibility of highly specialised staff, two consequences follow. The first is that we must be sure that the laboratory will continue to need that specialism in depth. The second is to consider the need for a system of payment under which salaries can be given to suitable specialist R&D workers comparable with those given to colleagues who have gone up the managerial hierarchy—the so-called dual ladder for promotion.

All of this underlines the need for managers to establish systematic practices for monitoring subordinates' progress, practices that will be both informal and formal. Most managers keep in adequate touch with their subordinates informally or through regular project reviews. This, though essential, is not enough. There is a key role for the annual (or two-yearly) formal appraisal review. Ideally this review is undertaken by an individual's immediate superior, and its purpose is to examine that individual's performance over the past year (or two) and from it to identify strengths and weaknesses, to define steps to build on the strengths and make more use of them, and to help overcome the weaknesses. Unless the interviewer is careful, it can relapse into a project review; it is in fact quite different. What we are after is to find out what the subordinate has done and how well, what skills have been used and how well, and in what direction the subordinate will go in the future. The emphasis is on performance and work behaviour, not personality.

For appraisal interviewing, many of the selection interviewing skills are relevant. However, whereas in a selection interview the interviewer will be evaluating the candidate, part of the purpose of an appraisal interview is to help the subordinate to appraise him or herself and to reach agreement between manager and subordinate about future performance targets. Through the appraisal interview (Lattram and Wexley, 1982) we agree on a set of steps to be taken by an individual or for him or her by the manager or the organisation during the next period. We consider enhanced responsibility, refreshed technical knowledge or skill, the possibility of attendance at courses on, for example, specific interpersonal skills, presentation skills, etc.

It is important to recognise that individuals develop at different rates, and to different ultimate levels. We need to enter into the way our subordinates see life: what their motivations and abilities are. Again, we need to be aware of the danger of forming too rigid a view of our staffs: people do change. The danger is that if we do form strong views, we are liable to view performance in a prejudiced way and to interpret it in the light of our already formed views.

## Capability

Research (initially in the Glacier Metal Co. Ltd, and then at Brunel University) (Jacques, 1977) has established that people differ in the ultimate extent to which they can handle complexity and in the time scales with which they are comfortable. But, more important, this capability is not a continuous spectrum but rather a series of discrete modes, in which the strategies used by the individuals are different at successive modes.

## Mode 1—time-span 0–3 months

This approach is characterised by the view that there is only one answer to the problem. The solution suggested will probably be directly related to the last object or event mentioned and very straightforward. If this solution is found not to be right the subject may well decide that the problem is insoluble. The approach used can be described as concrete and direct with no suppositions or abstract ideas involved. Reasoning in this mode creates an impression of working on a local or small scale immediately around the object or problem under review.

## Mode 2—time-span 3 months–1 year

This approach shows an awareness of cut-and-dried options to be chosen between. There is little possibility of overlap of these separate options or for compromise between them as the individual sees them as quite discrete. This is a very black-and-white approach.

The options are made up of positive information gleaned from the problem and are usually highly relevant to the solution but incomplete. The factors considered using this form of reasoning usually provide short-term or immediate solutions directly related to the current problem. The effects of the solution may be far reaching but this long-term aspect is rarely included in the choice between the options.

The overall impression of reasoning in this mode is that there is little cohesion between the different strands of the problem. This isolation makes evaluation very difficult and results in reluctance to discard an option as each option is known to be directly related to the problem. There is anxiety that the key to the solution may go with the discarded option.

## Mode 3—time-span 1–2 years

This approach looks for a way of linking individual situations into systems or patterns. Each item in the problem is linked into a sequence or chain of events until the solution is reached. If the series is broken the subject usually goes back to the beginning and starts again, even if only recapping what is known to date. While searching for the links that bind the solution together the individual is often inventive and sometimes uses 'gut-feel' or intuition.

Many items of information or experience that appear similar to the subject under discussion are drawn in while an indication of the scale and scope of the set is searched for. The approach appears outward looking but attempts to proceduralise and create order by finding a co-ordinating system.

## Mode 4—time-span 2–5 years

This approach uses the concepts of analysis of components and evaluation of the merits of two different approaches. The factors are teased out of the problem and two different theories of possible solutions are built up. The subject looks at both sides of the argument, weighing the pros and cons until one side emerges as more favourable. From that point on only supporting evidence for that side is considered. This is the first mode in which the use of negative information is included, and it is an important part of this approach.

The approach aims to structure material into a framework but can cope with gaps in the knowledge. These are seen like missing pieces of a jigsaw, defined in outline but not immediately available.

The overall impression of this approach is of one that spans a broad spectrum but focuses in great detail on certain aspects. It operates in abstracts and concepts rather than concrete issues but uses detailed facts and figures to support the ideas that are used to construct the overall framework.

## Mode 5—time-span 5–10 years

This approach takes an overview, searching for a relationship between apparently unrelated material. The solution could be described as holistic, where the whole is greater than the sum of the parts. The approach does not use detail to verify the solution but looks more to the gaps in the understanding as a source of inspiration for new links to be made to the existing areas of knowledge. There is no certainty that these gaps will be filled, but if they are it is the relationships between the new knowledge and the old rather than the parameters of the gap that determine the fit.

Negative information is used implicitly to exclude inessentials although these are not then totally discarded. The final solution is checked for misfit by a brief scan of all that is known to date. There is an expectation that the problem may change or be transformed in some way before the solution is reached. This facilitates long-term operating.

The overall impression of this approach is of an overview encompassing a large scale of activity. It is a flexible but complex approach looking for cohesive relationships that will meet the general criteria, and it operates over an extended time scale with details being devised and modified as necessary as the overall plan proceeds.

A number of practical consequences flow from these modes. For example, a mismatch between the level of the work to be done and the capability mode of the job-holder is almost invariably a source of trouble. If the level is too low, the job holder will be frustrated and bored, and will strive to find other ways of giving rein to his or her capability. This may take the form of voluntary work outside the laboratory or of an increasing involvement in professional organisations (or trades unions), to give two examples. (Note that the *level* of work is too low, not the quantity.) If the job-holder is subsequently promoted

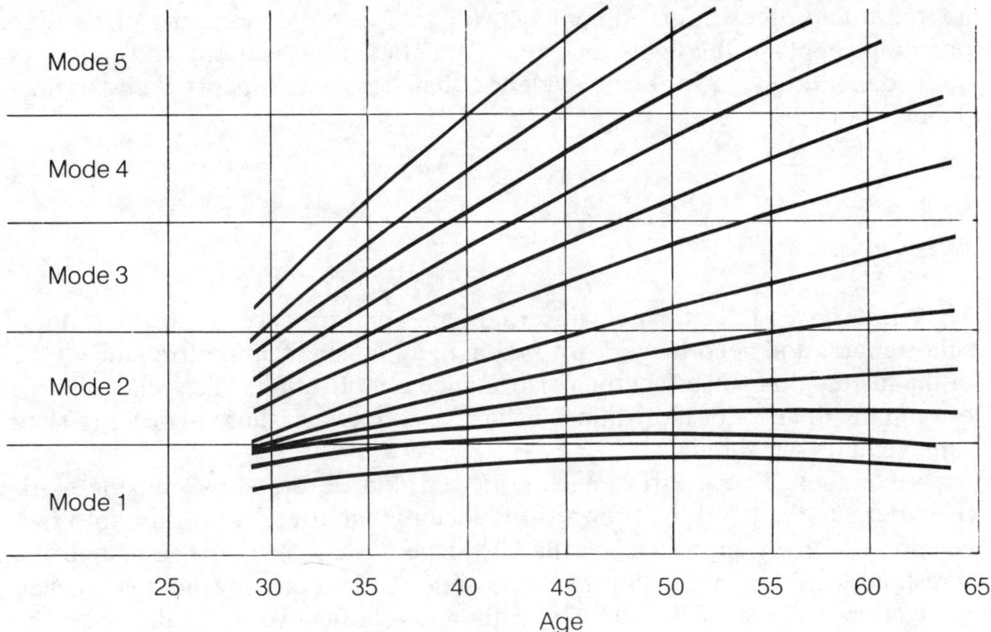


**Figure 1** Curves showing growth of capability with age.

to deal with an appropriate level of work, these outside activities will certainly decrease and probably stop, as the person's capability is fully used in work. If the level of work is too high, the job-holder will be anxious to an extent that may give rise to physical symptoms. He or she will be unable to do the task at the required level, will be unable effectively to delegate, will be submerged in the level to which he or she is accustomed, and the longer-term matters will not be done.

People instinctively expect to be managed by others of a higher capability mode. If the management structure has too many levels, they will go to considerable lengths to by-pass nominal managers whose capability mode is the same as their own.

By the time people are in their mid to late twenties, it is usually possible to judge the capability modc they will ultimately reach, and when. Figure 1 gives a typical growth curve. Some will not get beyond mode 2, some mode 3, and so on. Development is helped by giving people the opportunity, through the complexity and time scale of the work they do, to try out their capability and to match its growth against the extent of their accountability. Any individual is capable of using the strategies up to that of his or her own mode: the effective individual will select the strategy appropriate to the task in hand. When we communicate with others, we need to do so in a manner that matches their mode.

Finally, we need to beware of fixing labels on people and regarding them as correct for all time. We may be wrong, and people do develop. The annual appraisal and other informal contacts will give us the opportunity of revising our judgements in the light of current data. Indeed, it is useful for managers quite deliberately to look for evidence that does not support their original 'label'.

## *Motivation*

Managers have a legitimate concern in the performance at work of their subordinates and need to work with them in appraising that performance. One of the factors that will affect the performance is motivation. The social sciences offer many theories of motivation. I shall refer to two of these in order to shed some light on our subject.

In Herzberg's very early, much criticised, but nevertheless valuable work (Herzberg *et al.*, 1959), he grouped the factors that affect motivation into two categories. On group, which he called 'hygiene factors', need to be satisfied at a threshold level for motivation to be possible. Unless the threshold is reached motivation will be achieved only with great difficulty, if at all; once the threshold is achieved enhancement of hygiene factors will not motivate. The other group, which he called 'motivators', are those that, once the hygiene factors are satisfied, will actually motivate. Hygiene factors are broadly those matters that the organisation as a whole sets—pay, working conditions and environment, company policies, type of supervision. Motivators are much more in the gift of individual managers—an important point for them to realise. They include the content and challenge of the work; the actual achievements and performance of individuals; recognition of achievement and performance: opportunities for advancement.

Seen in this way, the organisation as a whole sets the platform on which motivation is or is not attainable and this role is crucial. Given this, the individual manager is able to provide motivators, taking account of the needs of the workpeople for whom he or she is accountable. Their motivational needs will differ. For some, for example, increased responsibility will be an effective motivator whereas there will be others for whom it would actually act as a demotivator. Nevertheless, within quite wide limits, managers can adjust the content of work and the consequent challenge it presents, and can devise ways of recognising achievement.

Expectancy theory (Vroom, 1964) is a more recent approach to the study of motivation. Here the concept is that for people to be motivated there is a chain of events that they must perceive: if they increase their effort at work, this must lead to an enhanced performance which must lead to appropriate rewards. It is people's perception that is important, for we all act on perception rather than fact. Two important factors emerge from this theory. One is the role of a manager as a remover of obstacles so that effort can result in performance. (It is surprising how many organisations have built-in procedures that militate against this!) The other is the need—as in Herzberg's theory—for a manager to know his staff as individuals and to be able to identify what rewards (some of which will be Herzberg motivators) will be meaningful to each of them.

In addition to opportunities already discussed above, those to attend conferences, to make overseas visits, to take the manager's place at meetings, to present work to senior staff or to important visitors can all be rewards for some people. Some will be helped by the opportunity to be included in the authorship of papers, by the chance to acquire a new skill, to be known as the expert in a particular technique or to help with the training of new staff.

Overall, of course, the interest the manager takes in the work and circumstances of staff is itself a considerable help in motivational terms, as long as it is not overdone and as long as it is done in a manner that recognises that individuals differ in the extent to which they need this type of encouragement and respond to it. We need to avoid the extremes of appearing to take people for granted and of complimenting them over-effusively on every minor achievement! There is a useful extension of this—a manager can let it be known that outstanding achievements have been made known to the manager at the level above his or her own. Again, this practice must not be overdone or, like the over-use of any motivational tool, it will lose its effectiveness.

There is one further point. The ability of the manager, the way in which he or she behaves at work and his or her achievements can in themselves act as motivators for subordinates. Managers who stick up for their teams, who are successful in obtaining resources, who can set up clear objectives, who are approachable and who are consistent and reliable are good to work for. Their teams are likely to be better motivated.

## The Team

I have spent some time dealing with some of the more important aspects of individuals at work. I shall now look at some aspects of team working, bearing in mind that practically all R & D is done by teams.

### *The research process*

Carlsson *et al.* (1976) developed a model of the research process from a model of the learning process by D. A. Kolb. This is outlined in Figure 2, and essentially suggests that research necessarily starts with what is known ('concrete experience'), moves through a period of 'divergence' in which leads are sought for a range of possible work, followed by 'assimilation', in which some sort of framework is formulated into which the alternatives can be fitted, and 'convergence', in which criteria are established against which success can be judged. Finally, we come to 'execution', in which resources are committed to specific (usually experimental) work, the outcome of which will be the

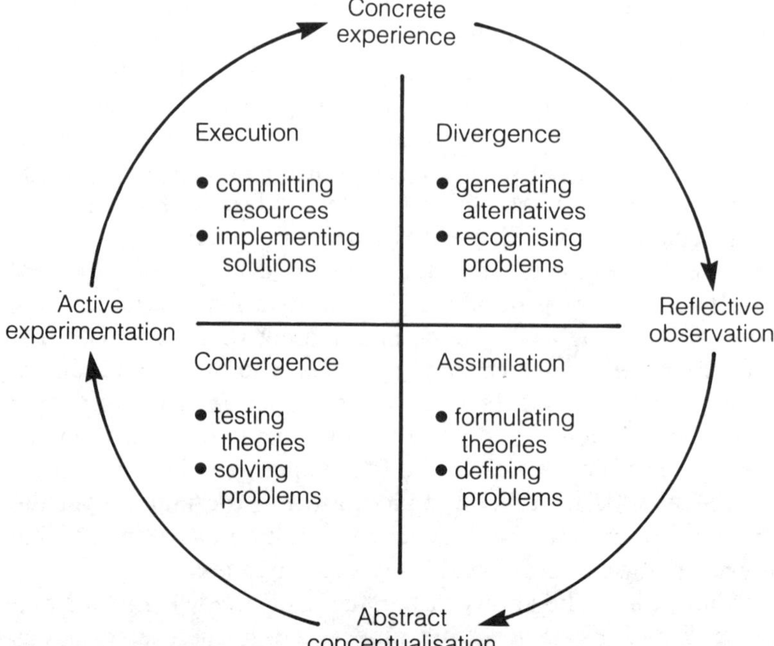


**Figure 2** Model of the research process (from Carlsson *et al.*, 1976).
Reprinted from *Sloan Management Review* by permission of the publisher. 

achievement of the objectives, or else an addition to 'concrete experience', on the basis of which the cycle may be followed round a second time and so on as necessary.

If this model is accepted we can see a number of distinct roles that will be necessary for effective working. The divergent phase calls for creative thinking, for the ability to produce novel ideas, for divergent thinking. This must be followed by the contribution that will critically evaluate these ideas and possibilities so that the most promising can be chosen for further work. At this point, the chosen idea must be turned into a practicable plan of action that must be carried through to a conclusion in which all the loose ends are effectively tied up. It will obviously help if the team can look outside itself and draw on work, resources and ideas outside itself. Of over-arching importance is the need for control of the whole process and, almost as important, the need to keep the team operating smoothly, to spot possible sources of conflict and to help the members to make the contributions of which they are capable.

A balanced team will be one that contains all these seven distinct roles, some of them combined in one person because we cannot accept seven as the minimum size of team.

At this stage, it is worth examining Figure 3, in which the consequences of excesses and deficiencies in the various phases are spelt out. Most readers will readily recognise situations like these within their own experience, and may well find that this lends credibility to the Carlsson model.

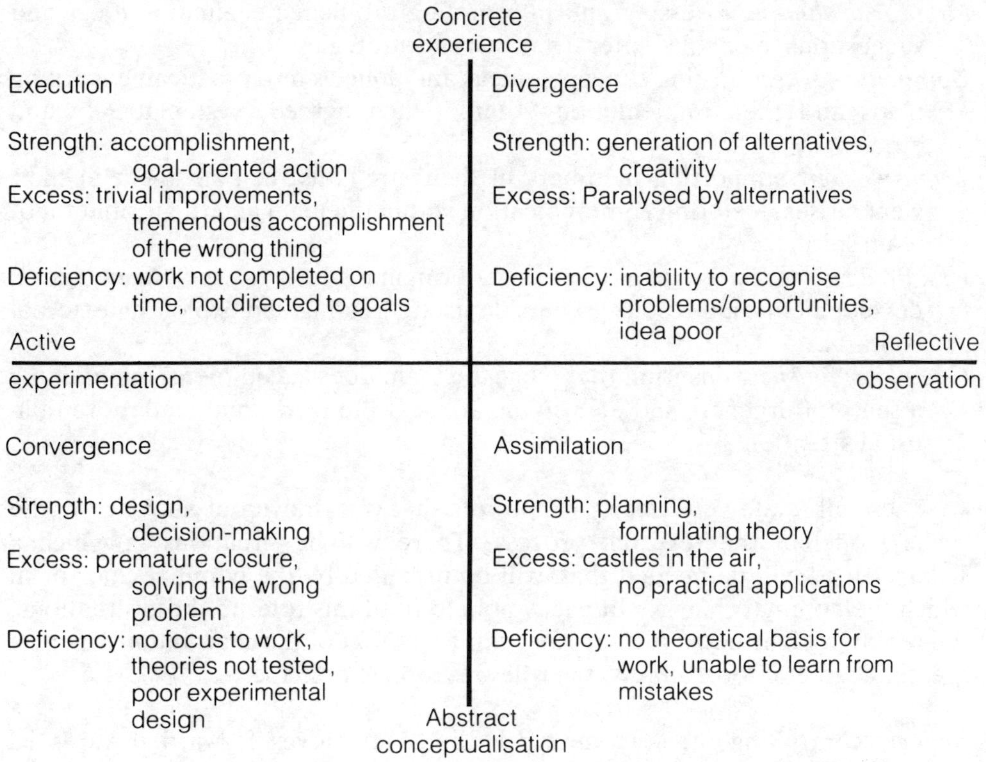

Concrete experience

| Active experimentation | Reflective observation |
|---|---|
| **Execution**<br>Strength: accomplishment, goal-oriented action<br>Excess: trivial improvements, tremendous accomplishment of the wrong thing<br>Deficiency: work not completed on time, not directed to goals | **Divergence**<br>Strength: generation of alternatives, creativity<br>Excess: Paralysed by alternatives<br>Deficiency: inability to recognise problems/opportunities, idea poor |
| **Convergence**<br>Strength: design, decision-making<br>Excess: premature closure, solving the wrong problem<br>Deficiency: no focus to work, theories not tested, poor experimental design | **Assimilation**<br>Strength: planning, formulating theory<br>Excess: castles in the air, no practical applications<br>Deficiency: no theoretical basis for work, unable to learn from mistakes |

Abstract conceptualisation

**Figure 3** Consequences of the model of the research process of Carlsson *et al.* (1976). Copyright as Figure 2.

## *Team roles*

It is fascinating to know that some entirely independent research into the constitution of successful teams, carried out by Belbin (1981), has resulted in the identification of what are essentially the same seven roles. To find such mutual support in quite different work is encouraging.

The research into teams was carried out by Belbin at the Henley Management College in England. The setting was six syndicates carrying out a business game with a measurable result. As the work developed and the various roles emerged, teams expected to be successful and teams expected to be less successful were deliberately formed and the predictions were on the whole fulfilled. The concepts have been used in real life, and again have been found to 'work'.

The roles are:

*Chair:* controlling the way in which a team moves forward towards its objectives by making the best use of team resources; recognising the team's strengths and weaknesses; facilitating decision-making.

*Plant:* advancing new ideas and strategies with special attention to major issues.

*Monitor evaluator:* assessing and interpreting problems; evaluating ideas and suggestions to enable better decisions and strategies.
*Company worker:* turning concepts, ideas and policies into practicable working plans and then implementing them (when agreed) systematically and efficiently.
*Team worker:* supporting members in their strengths, helping them in their weaknesses, fostering communication within the team and the formation of team spirit.
*Resource investigator:* exploring and bringing back to the team ideas, developments and resources outside itself; creating and exploiting external contacts.
*Completer/finisher:* ensuring that what the team does is complete, maintaining a sense of urgency, and identifying areas of the work that need more than usual attention.

These all relate very clearly to the role that was drawn out above from the cyclic model of the research process. There will be situations in which a stronger lead will be needed than will be provided by the nominal chair or in which the nominal chair is not easily able to fulfil this role. In these situations, the research identified a further role, similar to that of the chair but involving a greater degree of forcefulness: the role was termed a 'shaper'.

*Shaper:* controlling the way in which the team moves forward towards its objectives; shaping its activities and progress.

There was a final, ninth, role that emerged, that of specialist, concerned with the provision of knowledge, skill and know-how that the team needs at particular times. People whose predilection lies in this direction can, nevertheless, play one or more of the other roles.

What are the practical consequences of this research? Perhaps the first is in a sense negative. In R & D groups, the first consideration must always be to make sure that the team contains all the necessary theoretical and practical knowledge and all the necessary skills, or, if not needed continuously, that they can be provided *ad hoc* by specialists drawn in for the purpose. It will be rare to have the luxury of choosing team members on a team roles basis; scientific or technological knowledge must be paramount. In passing, it is worth remarking that in matrix organisations, in which teams are typically relatively frequently set up and re-formed, there may be more chance to allow team roles to play some part in the choice of team members.

Second, the performance of most teams can be enhanced by a conscious recognition of the need for a balance of roles within them. The team roles concept could well be presented to and discussed with the team. This can lead to two benefits. One is the realisation of the roles readily available within the team and the identification of gaps. Where there are gaps, we may well—and usually will—be able to identify a team member who, realising the need, can fulfil it. The second is the realisation that all roles are necessary and that the contribution of, for example, company workers (often, in R & D teams, the technicians who do the work at the bench) is essential for success; this can often be helpful for motivation.

Third, the role in which an individual feels most comfortable is a combination of personal characteristics and of recent work experience. Most people can reasonably readily perform at least adequately in two or three roles. Because it is rare for R & D teams to have as many as seven members, this flexibility is immensely important. If the team is aware of the need for a role in the team, a discussion can usually result in agreement on the team member most likely to be able to undertake it.

Fourth, in the case of vacancies at higher levels in an organisation, where people will be members of a management committee, the concept of team roles can be brought into the recruitment process. This will involve identifying the roles available in the management committee and then identifying gaps.

Fifth, work at Henley showed that team difficulties could be almost guaranteed to arise in certain circumstances. Thus a team with two shapers will almost certainly experience the conflicts that the presence of these two will engender: one will have to go or will have to withdraw from the role, and shapers find this very difficult indeed. Two plants in a team can give rise to conflicts as well, less pronounced than those engendered by two shapers, and usually resolved by one or other of the plants withdrawing from active participation. (Brain-storming, in which all participants are encouraged to behave as plants, is a different situation, artificially contrived in order to solve a particular problem, or arranged to cover the absence of an effective plant or to supplement his or her efforts. Even here, plants do frequently 'switch off'.)

## Team maturity

If we follow the progress of a team from the point of its formation, we shall find that there is an initial period in which team members are settling down, testing one another and establishing norms of behaviour, as well as getting down to the task for which the team has been formed. During this period, which in normal circumstances may last for up to six months, the natural roles of the members will become apparent, and the leader may well judge it right to discuss the research process model and the team roles within the team. Once the team has settled down, the work should go forward effectively, increasingly so for a couple of years or so. At this stage the leader (and even more so, the leader's manager) should begin to be aware of the possibility of staleness creeping in.

There is evidence to show that most teams that are left unchanged for too long—say four years or so—become stale. They show signs of self-sufficiency and self-satisfaction; they become resistant to fresh ideas from outside; they become introspective. There is therefore a need to introduce some change as a matter of deliberate policy. This will obviously be easier if the organisation as a whole has a policy of regular and systematic moves for the majority of staff. If this is the case, the moves can well act as motivators, bringing fresh challenges and opportunities to staff, and bringing the staff into a new range of human contacts. The extent to which various specialisms are required at depth will determine the extent to which this policy can realistically be pursued. The important thing is that moves should be viewed positively so that they are regarded as being in the interest of the individual.

## Notes

1. The chapter owes a great deal to the discussions that have taken place with approximately 500 participants on some 40 courses on the management of R & D, and with colleagues in Brunel Institute of Organisation and Social Studies, notably Dr Gillian Stamp, Dr Sheila Rossan and Mr Ian MacDonald.

## References

Belbin, R. M. (1986) *Management Teams: Why They Succeed or Fail*. London: Heinemann.

Carlsson, B., Keane, P. and Martin, J. B. (1981) R & D organisations as learning systems. *Sloan Management Review*, Spring.

Fletcher, J. (1986) *The Interview at Work*. Dolphin Puppeteers.

Herzberg, F. *et al.* (1959) *The Motivation to Work*. New York: Wiley.

Jacques, E. (1977) *A General Theory of Bureaucracy*. London: Heinemann.

Lattram, G. and Wexley, K. (1982) *Increasing Productivity through Performance Appraisal*. Wokingham: Addison-Wesley.

Vroom, V. H. (1964) *Work and Motivation*. New York: Wiley.

# 1.5 *Investment in Technology*

*Gordon Edge*

One definition of technology is 'the application of science for commercial value'. Given this, those involved in the management of technology are indirectly concerned in the management of science, transforming it through a combination of skills and resources into value added in a product, a process or a system. Managers are always concerned with the optimisation of resources whether they be physical, intellectual or fiscal, but the indirect involvement of technology management with science introduces a problem. This problem is one of uncertainty.

The conversion of scientific work into something that ultimately can be packaged and sold, usually with some kind of definition of performance, implies a high level of confidence in the underlying science to perform as we expect over a wide range of operating environments and conditions. In practice, this conversion process is undertaken by people, usually engineers, facing tough deadlines and cost targets within a rapidly changing competitive market. Their job becomes one of trying to optimise a prodigious range of compromises between an even larger range of variables. It is the rather indefinable nature of this optimisation process that gives a very special quality to the topic addressed in this chapter—decisions about investment in technology. In the early stages of a technological development these decisions are similar in nature to those taken when backing the writing and production of a new opera or play and, at the earliest phases of a new technological development, it is absurd to pretend that a rigorous method of assessing financial returns can exist. The nature of the decision then becomes one of experience and judgement based on difficult to define and subjective parameters.

In this chapter, in order better to understand the problem of technology investment decision-making, I will look at the nature, evaluation and accommodation of the risk being faced, before considering the reasons for such investments, investment decision-making and strategies, and sources of funding.

## The Nature of the Risk

### *Risk and uncertainty*

We frequently link risk and uncertainty, but it is useful to distinguish between them where investment in technology is concerned. Risk is involved where we can forecast a range of possible but quantifiable outcomes from our decision. On the other hand we are uncertain when we can find no way of stating how probable a particular outcome might be, if indeed it is possible at all. We cannot, as managers, back away from this type of uncertainty because this step frequently occurs in the early stages of any technological development and can lead to important competitive advantages in the ultimate product.

A scientist is usually involved in both uncertainty and risk during the early stages of his work. For example, he may be uncertain whether a particular piece of genetic manipulation will lead to the expression of a particular material. At a later stage in his work, although certain that expression can be achieved, he may still be unable to set the range of forecast production yields of that biochemical to better than, say, between 60% and 80%. This type of range statement gives rise to calculable risks of particular levels of financial returns and also to an indication of the justification for proceeding with that investment should yields fall below an economic threshold.

It is quite clear that the investment decision to be made at the earlier research stage is of quite a different quality from that needed when we are at the yield forecasting step. Put simply, we have moved from 'if' to 'how much' and this transition indicates the need for a considerable degree of flexibility on the part of the financial managers within a company. Note that this flexibility also needs to be coupled with some understanding of the scientific and technological issues concerned so that the financial manager can have some insight into the actual stage a development has reached.

### *Factors affecting risk*

Several studies have sought to establish the factors contributing to success in technological development work and I would like to quote from two, the first being Bergen (1983) and the second Cooper (1980).

Bergen's study compares UK and German performance in undertaking technological development programmes, but central to his theme are a number of hypotheses that draw upon an underlying statistical analysis. In summary:

- There is a strong correlation between R & D investment and company productivity (this was also indicated in an unpublished OECD report in 1984).
- Flexibility of operations and the absence of status consciousness within R & D and engineering development correlate strongly with all success factors.

- There is a correlation between the level of decision-taking and project success.
- Research content in the programme has a strong negative correlation with the adherence to project objectives. This is consistent with the introduction of uncertainty into the work as compared with predictable risks.
- Flexibility has the strongest correlation with success in many of the samples of technology development projects investigated by Bergen, particularly in respect of speed of development, which also has important implications for overall project cash flows and profitability.

In the case of Cooper, there was no direct relationship between size or quality of financial resources within a firm and the ultimate success of a development, but it is possible to infer that 'corporate strength', which includes financial and management resources, is a negative risk factor. This strength translates directly into strong structural and cultural linkages between the financial, marketing and technological resources within successful companies.

According to Cooper, some of the specific variables associated with new product success are:

- proficiently executing the launch (selling, marketing, distribution)
- better match to customer needs than competitor products
- having a higher-quality new product than competitors in terms of tighter specifications, durability, reliability, etc.
- undertaking a good prototype test of the product
- undertaking test marketing
- proficiently starting up production.

These studies provide a great deal of support for the contention that minimisation of risk in technological investment depends more upon the quality and culture of the technological skill and management resources than upon the methods used to evaluate an investment. The results also indicate the dangers inherent in making technology investment decisions depend too strongly upon the accountant's rather deterministic approach. Company boards dominated by accountants should take very special care to ensure that the technological areas of risk and uncertainty are properly communicated to them.

## The evaluation of risk

The evaluation of the risks involved in a technological development programme is in reality the aggregation of a large number of assessments, some objective but many subjective. This aggregation forms a major part of the process of 'due diligence', which today forms an inseparable constituent of the investment process.

Another new term has emerged in recent years, that of 'intelligent investor'. This type of investor is deemed to be knowledgeable of the subjective risks involved for the particular technologies to be developed and also capable of the formalised evaluation of financial risk. Furthermore, the intelligent investor will look for a much higher rate of return from his investment when the perceived technical risks are high. It is worthwhile, therefore, when considering any development, to partition it in terms of those phases of risk that are exceptional in terms of quality and of size. The early risks may depend upon pivotal experiments being undertaken, which in practice may not be costly and often may be financed internally from a general budget or from the inevitable 'skunk works'.

It will not always be the case that large cash reserves within a company enable larger risks to be taken. Sometimes, these cash 'mountains' have been amassed by not exposing the firm to technological risk and by making it in consequence a rather badly managed bank. Fortunately, shareholders have today become more sophisticated and expect technology-based businesses to make and manage risk effectively with consequently higher-than-average rewards.

## Due diligence

Due diligence is a highly descriptive term, lacking in precision, taken from the language of the professional financial investor. It is intended to convey the meaning that those proposing investments have exercised considerable care in the evaluation of the risks involved.

The first step in the due diligence process is a conscientious attempt to write down a pro forma financial plan for the proposed project. This plan will include:

- A phased plan of the research and development programme, including human and capital resources, time scales and some method of codifying the level of technical risk at each phase. Remember that software development estimations can be more difficult to produce than any other aspect of the project. Include the development of several prototypes at each stage in the development process.
- A phased plan of the transition from R & D to engineering and manufacturing development. Remember that 'design' encompasses the development from concept to completion and the involvement of conceptual, industrial and product designers in the team can raise the level of effectiveness considerably.
- A series of early estimates for capital expenditure in manufacturing technology, including the opportunities for advanced computer control and flexible manufacturing.
- Estimates of the cost of manufacturing the product at various volume levels.

- A phased plan for the development of marketing and sales and for the support of the product in the field.
- Estimates for the sales of the product from launch to perhaps three years out.

From this plan will emerge a very early estimation of the financial potential of the investment. Usually, the analyst will reduce this to a series of cash flows and make an initial estimate of two financial indicators principally needed to indicate choices between competing investments. The analyst will also wish to check the sensitivity of the investment to the many financial variables in the pro forma plan.

The most common financial indicators are NPV and IRR. *Net present value* (NPV) is an indicator of the current worth of an investment reduced to a series of cash flows. These cash flows will be taken from the business plan for the technology investment but since the plan covers a number of years it will be necessary to compensate for the fact that £1 spent or received three years from now will have quite a different value from today. The adjustment for the temporal shift in the value of money is made by discounting and the factor chosen for a particular investment is the discount rate. *Internal rate of return* (IRR) is the discount rate that must be applied to give a net present value of zero.

It is at this step in the preparation of a business plan that the first of what may become an interminable series of iterations occurs. It is unlikely that the first attempt at a pro forma business plan will yield an acceptable level of either NPV or IRR and, even if it does, the sensitivity of the financial return factors to individual parameters within the business plan must be determined. For example, what happens if the R & D programme takes longer or the sales build more slowly or the manufacturing facilities are delayed? This sensitivity analysis will indicate which plan parameters are most able to modulate the NPV and should suggest what has to be done to maximise its value.

It is important to avoid the tendency to construct 'hockey stick' business plans in which all the cash flows justifying the investment occur in the last few months of the final years of the plan. Look carefully for opportunities to create horizontal value within the plan:

- Can the technology investment be shared with others who might be able to benefit through exploiting the technology in non-competitive markets?
- Can any other part of the business benefit from commonality in investment?
- Can early opportunities be seen to sell the technologies to other organisations in a basic form—for example as a new materials technology?

The first method of the three is known as 'off-balance-sheet' financing because the incoming investment does not draw from shareholders' funds or reserves.

## Sensitivity analysis

When examining the sensitivity of the investment to various forms of risk, it is often useful to plot the 'probability' distribution of the cash flows rather than trying to select a single best estimate. For example, a common misconception is that the cost of manufacturing a mass-produced standard item always falls within an extremely narrow range. It is true that there is a high probability that the cost of producing a particular object, such as a car, will fall within this narrow range, but there is also a finite probability that single examples of the product might cost twice as much as an earlier example. This increase in cost might be due to some quality defect buried within the basic fabric of the car or a subsequent warranty claim might call for a new engine for that particular car.

The distribution of likely cash flows for a particular event in a technological investment will have a highly skewed distribution of probability. The skew will, inevitably, be towards increasing cost and time. Where R & D investments are concerned there is sometimes a finite probability that success will *never* be achieved at *any* cost. This phenomenon will be indicated on the probability distribution as a line asymptotic to infinite cost! The experimental programmes associated with one of these highly skewed distributions are called pivotal (because overall success 'pivots' around them) and should be undertaken before any major expense is committed.

These probabilistic cash flow distributions are usually best estimated by teams of individuals using their experience and judgement. Remember that the greater the standard deviation of the estimates, the greater the risk and the higher the rate of return that will be needed to justify the investment.

It is always helpful when assembling a business plan to involve a combination of financial, market and business specialists working together with the technologists so that the nature of the risks involved may be shared and understood. This approach minimises the emergence of confrontational positions when the plan is up for decision by a board or committee. Most chief executives (CEs) are more willing to take a greater risk than is the expectation put forward by those who seek to shield risks from them. This is, of course, provided that the CE can see that the risks involved have been reviewed and assessed in a thorough and professional manner.

## Risk management in small companies

The management situation within a start-up company will usually be quite different from that in the mature company, particularly as the latter will have adequate financial and management resources to deal with the inevitable problems that arise during the course of any technological development.

It is most important in companies of all sizes to ensure that the proposed project is adequately funded to cover the most likely anticipated level of risk. Few things preoccupy people in a start-up company more than having to re-finance a project when it runs into difficulties. At the very time when all attention should be concentrated upon resolving technical problems, valuable time is being lost seeking new finance, often under disadvantageous terms. The message is therefore to evaluate risk as thoroughly as possible before going for funding, whatever your resources, and then to devote your time to managing the implementation of the programme plan.

## Technology, Skill and Value Added

The principal reason for making an investment in technology is to raise value added within the product, process or service offered by a firm. This is because, in the limit, competitive advantage in the market place is controlled by value added; in particular because of the pricing flexibility a high and well-distributed value added can bring. A dominant value added position gives business managers the maximum potential for growth in market share and penetration with long-term benefits in terms of competitive advantage and profit.

This statement might seem difficult to justify until one remembers that value added can derive from the incorporation of skill *anywhere* within the full extent of a company's activities. This skill might, for example, be intangible and represented by a gifted salesman. More likely today is that the skill-based value added will derive from the activities of technologists within the firm and therefore advantage ultimately goes to the organisation that manages its technological resources more effectively than its competitors.

It is possible to develop a convincing argument for the association of skill intensity and value added over a very wide range of industries. Figure 1 shows one such relationship derived from MITI statistics. Skill intensity is the economists' way of describing the proportion of cost in a firm that derives from professional (i.e. educated, knowledge-based) skills. The term 'skill intensity' can be applied diversely and correctly to the skill base within a manufacturer of advanced optical glasses, within a supplier of financial services or within an accounting practice. This is rather an important observation because it gives an indication of future opportunities for inward investment in the technological development of service sector companies and in addition raises the number of options for ways in which manufacturers might improve their opportunities to compete.

The associated rising capital intensity within service sector companies will remove whatever perceptual differences remain between businesses in the manufacturing and service sectors. In both manufacturing and service companies professional skills fall generally into combinations of the following categories:

- research and development
- design
- production engineering

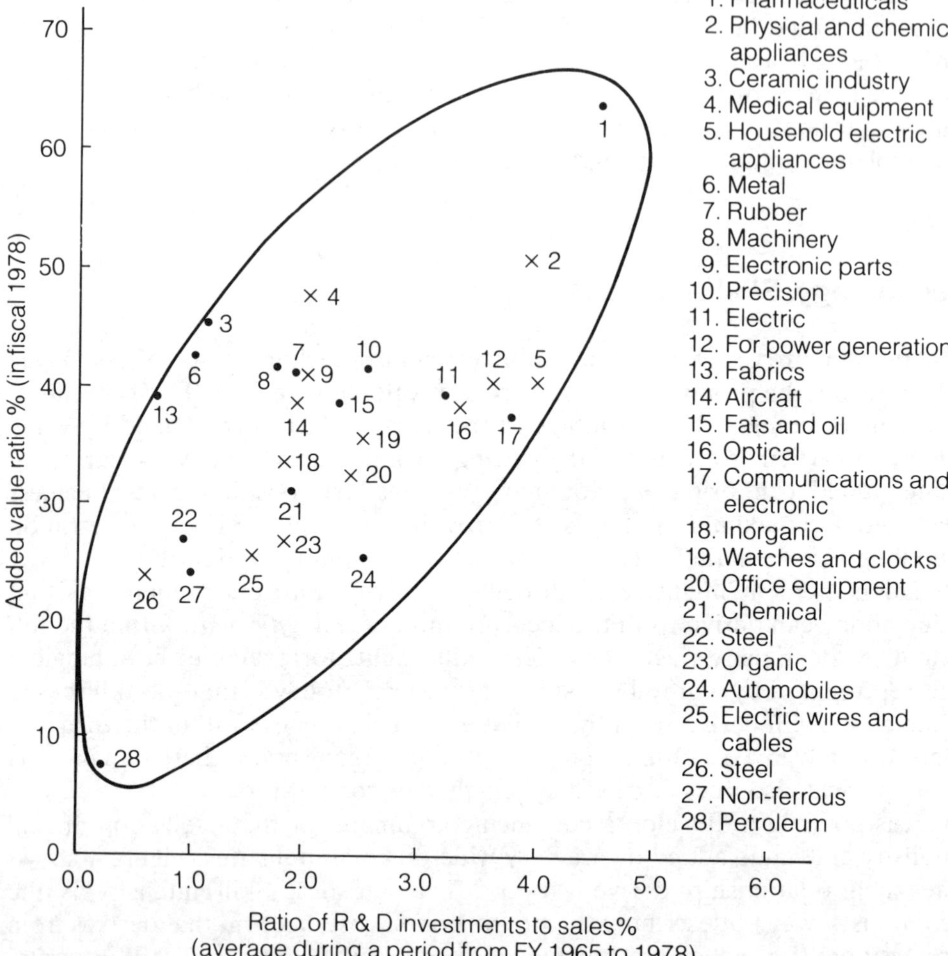


**Figure 1** R & D investments and added value ratio. Added value ratio = added value/shipped product value × 100% (added value and shipped product based on industrial statistics).

- manufacturing engineering
- financial
- marketing.

Furthermore, in most companies these skills represent a fixed cost and have associated with them the two important variables of effectiveness and efficiency. Of these two variables, one is qualitative and the other is quantitative. Effectiveness, the qualitative factor, is affected by the following:

- culture
- capital
- organisation

- environment
- quality of education and experience
- technology or knowledge base
- information (e.g. the management information system).

Efficiency represents the cost of delivering a particular level of effectiveness into the process or product (or service). It is particularly affected by capital intensity because proper investment in facilities to support expensive skill can have a dramatic effect upon effectiveness.

Both effectiveness and efficiency are affected by good management. For example, it is often better to spend more on a piece of laboratory capital equipment which is easier to use and calibrate and hence frees more skilled time for creative purposes, than to go for the lowest-price instrument. It is surprising how often this rather obvious point is overlooked.

In summary, therefore, while relative competitive advantage derives from relative value added, because skill is the correlate of value added it follows that, ultimately, competitive advantage derives from the relative efficiency and effectiveness of the skill base. Thus competition in international markets will depend more upon organisational and cultural effectiveness than access to low-cost, low-skill labour.

## Determination of an Investment Strategy

The determination of a technology investment strategy within a company should be based upon a self-consistent definition of the role of technology in that company, against which any future strategic choices or decisions might be referred. The basis might, as an example, be expressed in marketing terms so that the firm is perceived by its customers as a technological leader. Alternatively, the company might use technology to enable it to deliver a personal or individual design of a product, such as a gas seal, to a particular customer thereby gaining a differentiated service orientation. In any event the purpose of making an investment in technology must ultimately be financial and in this chapter value added (disaggregated) has been used as the principal purpose of technology investment.

This is because:

- Managers are primarily involved in value added rather than profit, which can aggregate from many sources unrelated to the business itself.
- Relative competitive advantage is a function of relative value added between competing products or processes.
- Value added is a direct correlate of skill intensity in a firm.
- Innovation derives from the effective use of skilled resources and is therefore within the process of strategic management.
- Value added maps across the entire organisation and hence any part of the skill base can contribute to it.
- The location of value added is itself a key strategic issue.

### *Technology strategy and value added*

The location of value added is not normally considered to be a strategic management issue but it is actually rather critical in managing investment in technological innovation to be sure of the target location of the value added as well as the amount. This is because the amount of added value in a particular location, such as materials or plating, tends to change with time and eventually appears to decline until it vanishes. The manager needs to be aware of and plan for this phenomenon.

As an example, take the case of the compact disc player. When first introduced this product located most of its value added-derived innovation in the technology of components such as the solid-state laser, VLSI chips and the disc itself. Indeed, the first generation of players was manufactured largely by hand. By the time of the third generation, little innovation was to be found in materials and product technology but a substantial amount of research and skill had gone into new manufacturing technology so that the players became smaller and more sophisticated, with a declining price profile. The value added provided at the product level had simply declined to be replaced by increased value added at the manufacturing stage.

This strategy of flexibility in continually reviewing and changing the location of value added is an essential aspect in managing a comprehensive approach to a technology investment strategy. The issues facing the manager investing in technology are:

- degree and location of value added
- skill source of the value added
- gearing between value added and investment
- flexibility.

## Organisation and Skills

Organisation, structure and skills are dealt with in detail in Chapters 1.4, 2.1 and 2.3, but it is necessary to make a few points here.

The role of the skill base within a company is to couple technology and knowledge base into the process, product or service that represents the business of the firm. This coupling must occur effectively and efficiently. The best type of structure to do this is one that is both multidisciplinary and interdisciplinary, remembering that the disciplines involved must extend beyond science and technology into the business-orientated skills of finance, marketing, information and so on. The structure must also recognise the strong vocational drives and needs of the professional while at the same time providing to customers and suppliers a commercial link that has integrity and quality.

The strategy and investment management function cannot be separated from these issues because it also has a contribution to make in the innovation process. The primary reason for this is the high level of iteration that occurs in an effective organisation with a continuous reappraisal of risk, goals and achievement. This is just as true in a start-up company as in an established manufacturer, although the close personal relationships within the start-up do simplify the iteration process. The large company must therefore ensure that the management structure and process support flexible, communicative decision-taking. There is also, by implication, a significant demand for the (re)training of technologists within any organisation in order to maintain effectiveness in times of changing technology and the nature of competitive markets.

This discussion points to a number of factors that will affect the success of a technology and investment strategy:

- that the technology organisation underpinning the strategy should be interdisciplinary, multidisciplinary and flexible, and extend throughout the business spectrum
- that the skill base should be of the highest possible quality and coupled to a technology and knowledge base and to the market in an efficient and effective way
- that the utilisation or effectiveness of the skill base is formally optimised by a management process
- that the corporate strategy and management resources should see themselves as a part of the technological innovation process and not as a remote supplier of cash.

## Practical Aspects of Technology Investment Strategy

The process of developing a technology strategy cannot be dissociated from the business strategy of the company itself and wherever possible an approach based upon the creation or identification of a portfolio of technologies should be used. (See also Chapter 1.3.) This will be initiated by the creation of a technology hierarchy for the company, resting upon a set of enabling technologies.

The progressive stages of analysis and synthesis in building the strategy will then be:

- audit of the existing technology and skill base
- mapping of technology-derived value added in selected divisions
- identification of gaps or anomalies in the value added chain
- construction of portfolios relating technology resources to market segments, contribution and market share
- analysis of the company business strategy to identify technological competitive determinants, such as competitor efficiency, possible technology discontinuities and supply side changes in materials

- synthesis of a generic map of the science and technology base orientated to current and longer-term needs of the company
- development of recommendations for an organisation structure able to maximise both the effectiveness and efficiency of the R & D process.

The strategy as a whole will then seek to orientate the technology investment programme so as to raise the value added in various parts of the company product, process or service chain.

## *Measures of technology investment*

The classical method is to compare aggregate R & D investment to sales within a particular industrial sector. Thus the average pharmaceutical company might spend 15% of sales on R & D while the typical mechanical products company might spend less than 3% of sales. These levels of investment might be judged in comparison with similar businesses in the same sector within particular economies.

This method of setting comparative ratings between R & D investments in companies, although used extensively throughout the world, has little value, because:

- The current investment in R & D rarely has any effect whatsoever on the current year's sales, but will probably impact between two and three years hence.
- Even if the comparison is extended out a number of years it is still no measure of the effectiveness of R & D.
- There is no standard allocation of cost to a standard definition of R & D.
- The R & D expense may be cross-subsidised from other R & D intensive activities, such as defence. Some major organisations manage to fund over 50% of their R & D in this way.

A more logical proposal is that historical R & D expense should be measured as a proportion of current value added. This approach, for the reasons discussed earlier in this chapter, measures actual expense and, perhaps more importantly, also measures R & D effectiveness. Interestingly, this type of measurement operates in reverse to the traditional ratio in that the more effective the R & D, the lower the proportion of R & D spending to value added. The measure is also applicable to service sector, skill-intensive companies such as software, design and consulting.

## Programme Evaluation Procedures and Criteria

Most effective and well-managed companies will have substantially greater demands on their financial and skill resources than can be fully met from within. There is, therefore, a need to set priorities and these will comprise a combination of formalised hurdles which any proposal must climb. In addition, there will be a series of decisions based upon the judgement and experience of senior managers, which in the end distinguishes the great from the good. Within this chapter we need concern ourselves only with the formalities!

The key principles are:

- programmes should be evaluated stringently for net financial benefits
- other qualitative or subjective criteria are recommended for consideration
- guidelines should be established for programme (and project) portfolio management
- programmes are evaluated and selected and priorities are allocated at board level for central funding, with inputs as requested from any level or function of personnel.

Such procedures and guidelines are absolutely required for the successful operation of a technology strategy.

## Risk evaluation

Most investment appraisal procedures assume the risk profile of the firm is not changed by any individual investment. With R & D this assumption may not be valid, and is unlikely to be valid when the proportion of a firm's resource devoted to R & D is changing. The two principal approaches to risk are increase in the discount rate, and risk portfolio management.

Variable discount rate is a reasonably common procedure, and is a more rigorous version of payback methods, which shorten required payback period according to perceived risk. Since variable discount rate as a method can get quite complex, it is probably as well to employ a simple approach, at least initially. Project risks can be grouped in a traditional way into (a) existing product enhancement, (b) new products into existing markets, (c) new products into new markets. These three groups will be found to need progressively higher increments to the discount rate. The actual increments, and short-cut methods, are discussed below.

Risk portfolio management requires the investing decision-makers to have full control over all investments within a company, and may not be practicable in a decentralised organisation. Strictly, the risk of an investment is defined as the variance of the expected outcome from that investment. The decision-makers will manage the risk–reward profile of their set of investments in a way that maximises the reward at any given risk. The performance of the company will then depend on the risk-preference of the decision-makers and on the availability of good projects at different risks. In practice, whereas it is entirely possible to analyse risk–return profiles historically, it is quite a different matter to use the method to direct future investment. The necessary analysis of risk can be handled more simply through sensitivity analysis, discussed overleaf.

### Life cycles

The discounted cash-flow method takes account of time in its net present value concept. But time is particularly important in development work in other ways. Elapsed time and development costs are closely related—strong correlation has been found, for instance, between project time overrun and cost overrun. In addition, the development time cycle tends to demand greater resources as it runs through its stages of exploration, feasibility, analysis and proof, product development, process development and testing. The weighting given to time sensitivity analysis must therefore be large, and will tend in development work to exceed that given to other normally important variables such as production cost and capital item expenditure.

Products also have their cycles, and an understanding of these is important in development work analysis. Thus development delays affect not only the expected discounted value of sales receipts, but may cause sales opportunities to be missed in part or in total.

### Terminal value

Often, a project analysis will depend nearly on the value of assets, IPR, market share and so on at the end of the forecast period. Where analysis seeks to justify capital investment, net book value of fixed and working capital is a reasonable and conservative terminal value to use. But in product development analysis, the primary asset created will be market position. While this has value, it cannot be taken as the net present value of future profits, as maintenance of value will depend not only on the injection of future capital, but also on further possibly risky development work, capital expenditure and so on. Where these are not known, or difficult to estimate, the net book value method should be used to prevent effective double-counting of future earning streams.

## The Preparation of Investment Proposals

Formal proposals must be written and submitted to those individuals within a company responsible for strategy, business development or other budgetary activity, such as the director of technology. Usually a programme will be supported by one or more of each of the following:

- chief executive
- R & D
- marketing
- financial planning

The proposal is written according to guidelines, and it should outline:

- a selection criteria format giving preliminary evaluation

- direct objectives and possible spin-off benefits
- predicted financial benefits to the company
- a clear programme plan showing critical paths, giving milestones with clear success criteria
- a clear assessment of technical and commercial risks, quantified where possible
- estimates of total costs: people, time, equipment, travel costs, legal, third party, patent
- an assessment of IRR with sensitivities
- analysis of skills required and identification of appropriate personnel, where possible (larger organisations will find useful an indexed skills database accessed in electronic form).

Programme vulnerability should in principle be contained by reducing overall programme time scales as much as possible—even if additional costs result—and by resolving issues that involve uncertainty as early as possible in the programme.

### *Proposal evaluation*

Senior individuals may ask one or more people to review the technical and commercial risks inherent in the proposed programme. They will identify unforeseen risks where possible, and may ask the proposers to modify the proposal. Existing programmes should be reviewed regularly against the original programme proposals with revisions agreed from time to time and used as a basis for further monitoring.

Some form of guillotine should be built into all programmes to be exercised when a series of preconditions for proceeding is not met. While this approach cannot be operated in a mechanical, unthinking way, it can represent a fine discipline both at the original planning stage and during the monitoring process. For example, existing programmes that miss phase milestones by more than, say, 10% (in terms of either time or cost) are reviewed immediately; new, modified proposals are submitted if appropriate.

## Programme Selection Criteria

All programmes (and project proposals) should be selected against a set of criteria. A sophisticated, but easy-to-use, format should be developed. Some general suggestions are made below.

### *Financial criteria*

A recommended method for the evaluation and comparison of proposals is the net present value of the incremental cash flow at the risk-adjusted hurdle rate.[1] Various short-cut methods are available and may be useful at early stages of project screening, or so-called precompetitive development.

Discounted payback and payback[2] are little quicker than NPV and are not recommended. It is better to turn the calculation around and to determine the benefit required to justify the expenditure, then to check whether this benefit will be available, with a large margin to spare. Fully quantified, this method gives the benefit-to-cost ratio[3] used to rank projects under the NPV method. A related method is the sales-to-development ratio,[4] which has been shown to correlate well with IRR, and which compensates for its rough and ready nature by requiring few assumptions in its calculation.

The following factors should be taken into account in such evaluation.

## Marginal analysis

Projects should be analysed on what they are expected to change: what extra costs they will cause, what extra benefits will arise. In principle, a firm should rank its projects by net benefit, and accept all those for which net benefit is positive. In practice, resource constraints and risk adjustments will mean that many positive but lower-ranked projects will not be indulged.

Projects that share costs or benefits are analysed in one of two ways. Either one project is a necessary precursor to the others, in which case it is analysed on its own, and the marginal analysis of the others assumes the existence of the first; or the projects are genuinely interlinked and must be analysed as one larger project.

## Marginal costs

Marginal costs to the company arise through the employment of extra staff related to the project, extra external costs and fees (such as those of outside experts), all extra equipment (at full capital cost, not merely extra depreciation), extra travel, accommodation, consumables, and so on. They also include the opportunity costs of benefits foregone: the employment of sales engineers on projects, diverted management resource, and so on.

## Marginal benefits

Benefits ultimately derive from the market place. Product innovation, product cost reduction, step volume increases in product, increased service efficiency and so on can be related to revenue in ways analogous to the traditional gain or defence of market share. The uncertainty of what may be projections a long way into the future is handled by a combination of risk compensation through the discount rate and sensitivity analysis. Thus, even if a project is expressly designed for cost reduction, analysis must consider the potential market benefits from a lower product cost, rather than merely assuming increased margins.

We should note that:

- R & D investment is rarely evaluated stringently, but it should be.
- Scientists and engineers, and indeed accountants, assume that it is not possible to evaluate R & D benefits through contribution to value added. It is, but there is a large confidence and attitude barrier to be overcome by almost everyone in accepting this evaluation procedure.
- For long-term strategic programmes it will not be possible to estimate precise costs and benefits. The exercise of forecasting these is, however, of enormous value, in terms of attitude and applying thought to these issues. Probability/sensitivity/risk evaluation of the forecasts can productively be applied to reach a consensus evaluation.

## Sources of incremental cash flow

The business activity within a company intended to benefit from the proposed expenditure must have a logical boundary put round it, and estimates must be made of the cash flows across this boundary under a set of premises representing existing business conditions. The business is then changed by adding the expenditure being analysed, and the cash flows are re-estimated using assumptions revised only where the expenditure justifies it. The difference between the flows is attributable to the investment.

Incremental revenues arise typically from sales, possibly from licences, and possibly from maintenance of sales which might, but for the present project, have been lost. Incremental costs can conveniently be split into fixed and variable, the latter varying with sales volumes. Fixed costs will only be incremental if they depend in some way on the existence of the project. Incremental costs should always be analysed to exclude purely allocated costs. Incremental savings should be analysed as for costs. They need to be well justified, and are either a reduction in costs presently being incurred, or a reduction in costs that will be incurred if the project is not pursued. Opportunity costs and savings should be demonstrated as the cost or saving of the next best alternative, and should always be included. Other, non-operating, cash flows arise from capital expenditure, investments, building of working capital, taxes and terminal value.

## Non-sources of incremental cash flow

Depreciation is not a cash cost, and should be added back if incremental depreciation expense is charged. Depreciation does, however, reduce tax charges and its effect should be shown under this heading.

The investment decision and the financing decision are normally independent of each other, so that it is inappropriate to include specific financing charges, such as interest on debt, in the calculated incremental cash flow. For the same reason, all capital and investment expenditures must be assumed to be made for cash. Leasing costs and repayments should not be substituted, as they include implicit financing decisions and corresponding charges.

### Inflation

Inflation affects not only unit costs and prices, but also capital costs, working capital and taxation. It must be expressly included, preferably allowing for relative inflation changes between different cost and revenue items.

## *Qualitative criteria*

Qualitative criteria for use in programme selection include strategic needs, which can generally be separated into two aspects: overall group strategy; and individual company needs. The technology strategy group, where it exists within a company, should be able to evaluate both aspects in terms of changing markets and technologies. Company CEs and directors of technology should have input in terms of their strategies.

The interdisciplinary approach is also included. Does the proposed programme require multidisciplinary skills and is an interdisciplinary approach proposed? Does the proposed programme use technology resources in a productive way?

Finally, there is the subjective impression of potential. Although this criterion needs to be treated very critically, it is valid. If the programme requires basic research that cannot yet be said to have guaranteed strategic implications, and if benefits cannot yet be fully appreciated, the number of people (named) at whatever level, in whatever function, who are prepared to champion the programme should be taken into account.

The programme proposal should contain a criteria format with input on all criteria. The proposal team have thus carried out their own evaluation and selection, are committed to the proposal, and save time for board consideration. Priorities will be set according to the above criteria. Personnel requirements from operating companies' proposed programmes will have been stated clearly in proposals. The company director/manager of technology will have approved this use of personnel by this stage.

The group director of technology, or his or her equivalent in a business, should have discretion to spend a reasonable sum for early risk assessment projects in relation to their potential to support major strategic innovations or emerging technologies. Discretion for a number of such projects each year is reasonable. Projects that are estimated to cost ultimately more than, say, £50 000 should all be subject to strict evaluation; even small, first-phase (£50 000 to £100 000) ones. It is not unusual to find that poorly assessed projects starting out with a relatively low budget end up with a significant long-term cash drain. The risk is certainly not reduced with lower anticipated cost!

It is a valuable exercise when long-term projects are undertaken to establish an independent review team. This group should be drawn from peers of those directly involved in the project so that views can be exchanged quite openly. The review team should include some non-technologists.

Not only can programme review teams play a role in ensuring effective programme management and operation, but the review team manager and project manager should be jointly responsible for following up and quantifying the success of the programme retrospectively in terms of financial (and strategic) benefits. They should report to the board or similar authority on the completion of each phase. An overview of collected retrospective reports will be one basis for reassessment of structure, organisation, strategy and management of company resources.

A knowledge of the true cost of skill and capital within the technology resources of an organisation is an essential underpinning to the technology investment and management process. The actual categories of allocation of cost will depend upon the nature of the parent company but as a guideline, the following categories are typical:

- employment costs (include salaries, social and other benefits)
    - professional
    - research assistant
    - support staff
- premises (include building lease, insurances, maintenance, etc.; stationery and post)
- communications (telephone, etc.)
- recruitment
- training
- legal and patenting
- computer software maintenance
- consumables
- third party costs (e.g. consultants)
- depreciation (or whatever basis the company uses to assess the cost of capital).

All these costs may be used to determine the equivalent daily rate for those involved in undertaking technology programmes within the company.

Remember to allow sufficient time for holidays, sickness and training when calculating equivalent cost (as a guideline use 64% as the average utilisation of professional staff). Remember also that projects will of themselves attract additional costs through special capital expenditure.

The budget will be used in a number of ways:

- as the basis for estimating costs in development proposals
- as the basis for managing costs during the year
- as the means for determining the cost associated with growth through recruitment to support the wider aims of the parent company
- as an audit base for work undertaken for third parties (e.g. defence work).

Some companies adopt a budget roll, where the budgets for the business as a whole are reassessed in the light of performance on, say, a quarterly basis, with adjustments being made to correct deviations during the remainder of the budgetary year.

It is in the interests of the manager of technology resources to develop for him or herself a realistic budgetary management system, even where the parent provides a strong level of control. This is because arbitrary allocations of overheads and other costs can be made by the centre, and give an unrealistic view of the true cost of R & D. This has led to a situation where technologists have a naive view of R & D costs, which reinforces the rather ambivalent attitudes that some company boards have to the financial management capabilities of their technological staff.

In the event that technologists decide to leave such companies to set up their own businesses, inexperience in financial management can leave them vulnerable to predatory investors.

## The Funding of Technological Development

In general *capital investment* decisions associated with technological development will be considered to be long-term and the estimation of the cost of capital to a company under these conditions is by no means straightforward.

The cost of capital is not simply determined from the price (e.g. interest) paid for it because in allocating funds to a project in competition with another, a company must take into account opportunities lost to it by tying up capital (the opportunity cost). On the other hand there is no universally agreed method of calculating the cost of capital for companies, even in the same sector.

In a theoretical treatment of capital budgeting, the assumption is that those projects having a positive net present value should be accepted for investment—this calculation being made after discounting the revenue and cost streams at the marginal cost of capital to the firm, as has been discussed above. One writer suggests that the cost of capital is the rate of return a company must earn on an investment in order to maintain the value of the company. The actual calculation of the cost of each source of capital will be completely dependent upon the circumstances of the project and the company seeking finance.

It is well to remember that, probably, the principal duty of the financial manager is to minimise the cost of capital within the company and that he or she will traditionally use a combination of equity and long- and short-term debt (gearing) to achieve that. New, more sophisticated ways are now available to enable the hard-pressed finance directors to use sources of capital that have less impact on the balance sheet, although of course the positive contribution from successful projects supported by off-balance sheet financing might also be less.

### *Equity*

Investment in any project will represent a call on the shareholders' funds because of the profit and retained earnings sacrificed in the medium term. Of course, in practice, the technology resources within a company represent a fixed cost and in a sense it makes little difference to the profits whether a project is started or not, unless the R & D resources are physically removed from the company. Indeed, the plundering of R & D resources has been used as a device to bolster short-term profits in companies that have been the subject of hostile takeovers, but not without a cavalier disregard to long-term security of the firm.

It is important, therefore, for the technology managers within a company to ensure that they have considered, in a flexible manner, alternative approaches to a project in terms of their longer-term implications on the cash flows, value added and effect on balance sheet reserves. Where a company has a high level of debt then adverse changes in the debt equity ratios may well be a consequence of unforeseen problems in a technological development that can then make a company vulnerable to takeover. On the other hand, a successful technological development can make a disproportionate contribution to the long-term integrity of a company.

## *Venture capital*

Traditionally, financing technology projects at their time of highest risk and uncertainty is known as venture capital. In practice, the venture capitalist exercises the same disciplines in the evaluation of projects as does the traditional firm but may be prepared to accept a higher level of risk to shareholders' funds than a mature company with a history of caution.

The venture capitalist shelters his or her risk by setting a very high price on the investment and creating a broad portfolio of investments, usually across a number of market sectors. He or she may also syndicate the risk with a number of other venture capitalists. The due diligence process can provide sheltering of risk by obtaining professional opinions on the many aspects of a proposal that may be challenged later in court if the venture fails.

It is in the interests of the investee to ensure that his or her internal standards of due diligence are higher than those of the investors and the minimum number of syndicate members is involved in the appraisal until the proposal is free from inaccuracies and inconsistencies. The venture capitalist will tend to push you into a minimum investment to meet that plan and therefore, as was discussed earlier, you should be quite sure that you have provided for adequate contingencies and that you are very thorough in the evaluation of all risk factors.

## *Off-balance-sheet financing*

In recent years the increasing costs and risks of technology developments has created new forms of financing that are designed to minimise their impact upon shareholders' funds. This form of funding is known as off-balance-sheet financing and can involve:

- external sources of venture capital
- external funding such as government or federal

- other companies
- technology transfer agencies.

In practice, the investment is made so that the participants benefit from success in ways that are non-competitive—for example, through a licence to use the technology in a different market or from sharing of revenues through a royalty.

The vehicle for the development can be a direct investment in the skill resources and facilities but this necessitates auditable R & D costs within the firm. Another method is the creation of the development company that is authorised to finance and exploit the technology but may have separate management control from the direct participants. Universities and other institutions can participate in these companies and obtain a longer term benefit through royalty flows and in some cases capital gains if the development company is subsequently purchased. It is also possible for the source of skill to obtain a 'carried interest' in which the intellectual property rights (IPR) are exchanged for equity.

This is an interesting illustration of the underlying interchangeability of intellectual and fiscal capital and one that I anticipate we shall hear much more of in years to come as the capital equivalence of skill and knowledge are recognised in technology investments.

## Notes

1. The 'hurdle rate' is the discounting rate used by a firm, and is so-called because of its use in IRR screening.
2. Payback is the time taken for a cumulative net cash flow to turn positive, i.e. the elapsed time before the proposed investment has paid for itself. Discounted payback is similar to payback but uses discounted cash flows instead of nominal cash flows.
3. Most projects incur their costs in the early stages, with benefits coming later. These, discounted, can be expressed as a ratio to show the quality of the project. This method is less appropriate if there is an overlap between costs and benefits, or if reinvestment is required.
4. $$\text{Ratio} = \frac{S[(1+g)^{t-T}-1]}{D\,g(1+i)^{t}}$$

   where $S$ is first-year sales, $g$ the growth rate in sales, $T$ the delay between development and first-year sales, $t$ time, $D$ the development costs. The ratio should be used principally for early screening, with a cut-off value around 10.

## References

Bergen, S. A. (1983) *Productivity and the R & D/Production Interface*. London: Gower Press.

Cooper, R. G. (1980) *Project Newprod*. Quebec Industrial Innovation Center.

# PART 2

# *The Impact of Technology on Management and Organisation*

The two chapters in this part of the book consider the relationship and impact of new technology on organisational structures, and the impact of technology on management.

The first chapter takes the 'macro' view in examining the relationship between the nature and level of technology within the organisation, and the structure, 'shape' and nature of that organisation. It will be shown that, often, technology and organisational factors are related. Some consideration will be given to the reason for this relationship, but the emphasis in this chapter is upon identification of the practical implications.

The second chapter deals with management roles and jobs. The impact of new technology on responsibilities, structures, relationships, etc. will be examined in order better to understand the types of approach, attitudes and skills required of managers operating in a technology-dominated industrial or commercial environment.

# 2.1 *Technology and Organisation Structures*

*Ray Wild*

There is often a relationship between the type of technology associated with an organisation and the structure of that organisation. The reasons for this and its effects are matters of debate, but its existence is accepted. It is this relationship that I shall consider in this chapter. It will be appropriate to start by considering some terminology.

*Technology*. Here, I shall refer to process- and product/service-related technologies, i.e. the hardware and software employed by an organisation or incorporated as part of its products or services. I shall focus on the new, in particular information-related, technologies.

*Organisation*. I shall deal largely with the *structure* of organisations—whether at the level of the firm, or some subdivision, e.g. department. Such organisations comprise people, in roles, brought together for the purpose of achieving specific objectives. We will concentrate mainly on formal organisations—put together deliberately and with some degree of permanence.

## The Nature and Importance of the Technology–Organisation Relationship

If an organisation makes a 'step-change' in the type of technology it uses, it is likely that there will be organisational consequences. For example, if a manufacturing company implements computer-aided design (CAD) and computer-aided manufacturing (CAM) certain organisational changes may become *desirable*. It may be appropriate formally to change the roles and relationships of designers, draughtsmen, process planners, jig and tool planners and machine setters. Whereas previously there may have existed separate departments for design and for process engineering with responsibility for machine setting resting with production supervisors, some combination may now be desirable since the machine setting task is virtually eliminated, and the same computer database serves product design and process engineering. In addition these technological changes make *possible* other organisational changes. It may no longer be necessary for those involved in design to be physically close to those responsible for production, and a more decentralised structure is possible. This simple case illustrates a form of technological 'push', i.e. the introduction of new technology giving rise to some organisational changes and facilitating others. This is one view of the technology–organisation relationship, often referred to as the 'technological imperative'.

While some companies may introduce new technologies simply because they are available and in use elsewhere, and/or introduce them for learning purposes, many have other motives (Child, 1984). Major technology changes, like the above case, will often be prompted by market, demand and competition factors. For example, the desire to improve product quality, to be able to offer a wider range of products, and to provide such products with less delay or lead time, have been among the major manufacturing goals of many companies in recent years. Such policies have often been adopted in response to market and competition changes. The introduction of new technology, e.g. CAD/CAM, flexible manufacturing systems (FMS) and computer-integrated manufacturing (CIM), is a means to achieve such goals. Given this perspective, therefore, technology can be seen as an intermediate variable. Certainly the same organisational changes might result from technological change, but additionally organisation changes may result directly from the adoption of the new manufacturing policies. Indeed, in this type of situation, there may be greater motivation for the company to implement some of the organisational changes that are made feasible by the new technology.

Further, we might identify a situation in which a company, having adopted new policies, restructures its organisation and then seeks to introduce appropriate technology to support this change. For example, a financial service company may adopt an objective of spreading its activities over a larger

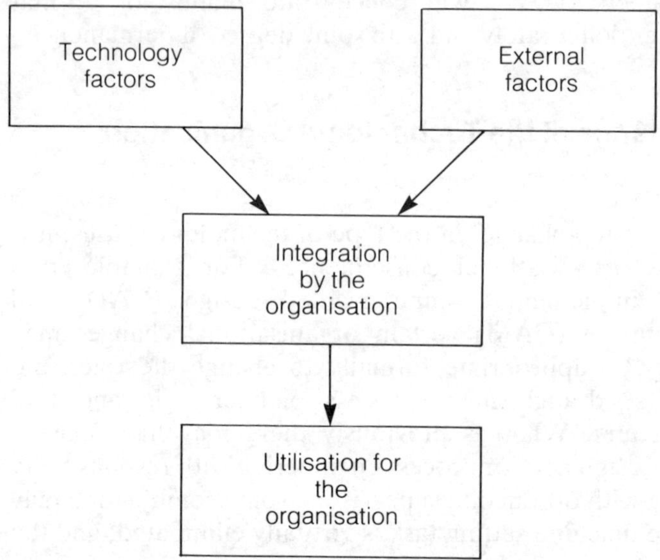


**Figure 1** The management of technology and external factors.

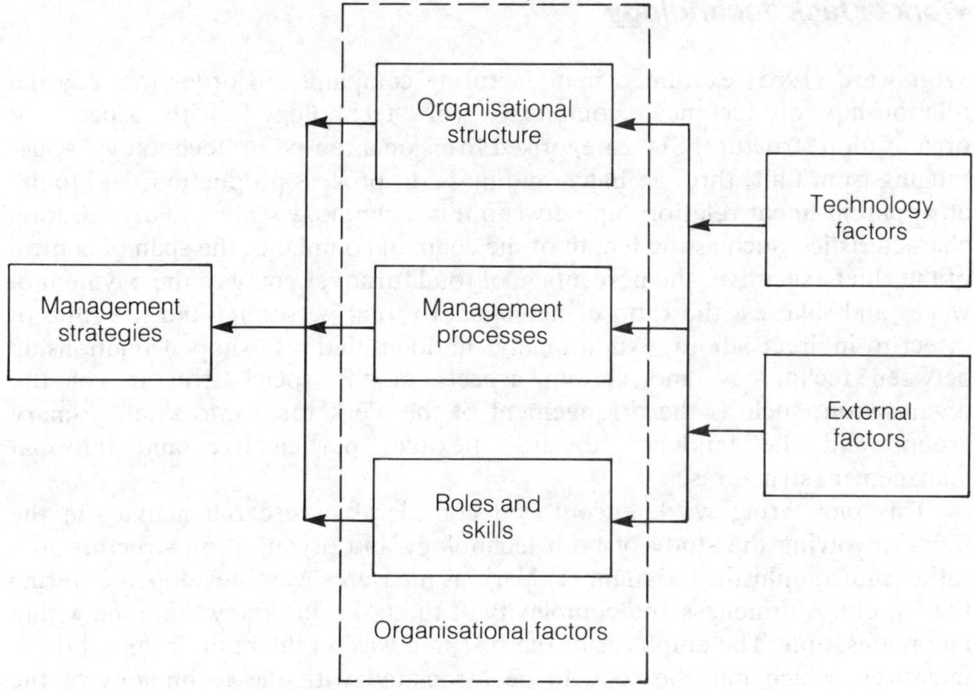


**Figure 2** The matching of organisational and technology/external factors.

number of smaller offices, in order to achieve closer customer contact. This decentralisation/distribution of the organisation may necessitate the adoption of new information and communications technology.

While it is necessary to have some view of the possible mechanisms linking technology and organisation factors, the important aspect for us is the outcome. In fact, since ours is a managerial perspective we can best see the issue in the manner shown in Figure 1. Clearly, there is a need for managers to deal with the increasingly turbulent, external environment, to utilise new technology and to accommodate or integrate these factors within the organisation, for its benefit. One management task therefore is to do with matching. Figure 2 presents a similar picture—again emphasising the need for the organisation to 'fit' with the technology and external factors to make it possible to pursue effective strategies.

This chapter deals with this concept of fit, or the process of matching. We concentrate on the structure of the organisation. The next chapter deals with some of the other organisational factors identified in Figure 2.

## The Technological–Organisation 'Fit'

It may be helpful, now, to review some of the major studies that are relevant to this area.

## *Work or task 'technology'*

Woodward (1965) examined manufacturing companies in order to study the relationship of technical complexity (or 'technology') with aspects of organisation structure. She categorised firms on a ten-point 'technology' scale, ranging from unit, through batch and mass, to process production, and found an apparent linear relationship between this technology scale and organisation characteristics, such as the length of the chain of command, the span of control of the chief executive, the percentage of total turnover going to the payment of wages and salaries, the ratio of managers to total personnel and the ratio of direct to indirect labour. Additionally she identified a U-shaped relationship between technology and certain aspects of the social structure of the organisation, such as the arrangement of the work force into small primary groups and the tendency towards flexible, participative and informal management structures.

This pioneering work encouraged considerable research activity in the field—involving the study of both technology and organisation structure in a rather more sophisticated manner. Various measures were developed covering the variety, routineness and complexity of the tasks and the work done within an organisation. The emphasis of the research was on the relationship of these measures, which may be seen to be associated with the technology of the system, and aspects of organisation structure. The two sets of factors studied are outlined in Table 1. The principal conclusions obtained are summarised as follows:

1. *Task uncertainty*. The greater the routineness or repetitiveness of tasks and the less the task variability, complexity and uncertainty, then the less the degree of participation in organisation decision-making and the greater formalisation of roles, procedures and practices.
2. *Task interdependence*. The greater the interdependence of tasks, roles and activities and the less the rigidity of workflows, then the greater the participation in decision-making and the less formalised the authority structures, procedures, etc.
3. *Workflow uncertainty*. The less the variability, complexity or uncertainty of workflows, the greater the standardisation of inputs and outputs, then the more formalised and centralised the management, the greater the vertical integration and departmentalisation, and the more sophisticated the control procedures.

   A further conclusion, while not primarily related to technology as we define it, should be noted.
4. *External uncertainty*. The less the rate of change in product/service specifications and ranges, the less the rate of programme/demand change, the less the market uncertainty and the greater its homogeneity, then the more mechanistic the management of the organisation, the greater the formalisation and centralisation of management, and the more structured the organisation.

**Table 1** 'Technology' and 'organisation' factors or measures

| Technology factors (relating mainly to the nature of the work or tasks performed within the organisation) | Organisational factors (relating to the nature and shape of the organisation and the manner in which decisions are made) |
|---|---|
| *Task uncertainty*, i.e.<br>(a) task difficulty, e.g. the extent to which there are known and adequate procedures and adequate time for tasks<br>(b) task non-routineness, e.g. the amount of variety in and/or the number of changes in the task over a period of time<br>(c) task unmanageability, e.g. the extent to which the task and any changes in it are understood by the worker | *Specialisation*, i.e. to the division of labour within the organisation, e.g. the number of specialisms and functions, the degree of 'role' specialisation, the specificity of tasks<br>*Standardisation*, e.g. standardisation of procedures, the extent of rules or definitions, standardisation of roles |
| *Task interdependence*, e.g. the need for active self-initiated collaboration and co-operation between workers | *Centralisation*, i.e. the focus of authority to make decisions affecting the organisation, whether formal/institutional authority or personal authority stemming from knowledge and experience, e.g. the location of decision-making, the promulgation of rules to limit discretion, review procedures and control systems, and the availability of information |
| *Workflow uncertainty*, i.e.<br>(a) workflow variability, e.g. the number of possible or alternative sequences of activities available for each item<br>(b) workflow complexity, e.g. the difficulty of determining the appropriate sequence or next stage of activity for each item | *Configuration*, i.e. the relationships between positions or jobs, described in terms of authority and responsibility, e.g. vertical and lateral spans of control and numbers of positions (or jobs) in various segments, etc. |
| *External uncertainty*, e.g. the rate of change of product or service characteristics, or demand, and the heterogeneity of demand | |

Taken together these conclusions suggest a negative relationship between the extent of uncertainty in (1, 2 and 3 above) and around (4 above) an organisation and the degree of structuring and concentration of authority and decision-making of that organisation. We might expect this type of relationship to exist at different levels within the organisation. For example, the uncertainty of tasks, their interdependence, the uncertainty of workflows and the demands placed on a department and the workgroups within that department will be reflected in the organisation of both the department and its workgroups. At the organisational level, a high degree of structuring and concentration may be evident through functional divisions, staff/line relationships, reporting and control mechanisms, etc., while at the workgroup level a similar structure or style may be evident through detailed job descriptions, work procedures, closeness of supervision and high division of labour. At the organisational level, low structuring and concentration may be evident through the use of 'matrix'-type structures, project or task group arrangement, working parties, an emphasis on informal communications, frequent role or job changes and transfers. At the workgroup level such an approach may be evident in job rotation, worker control of workflow, absence of formal supervision, the use of semi-autonomous workgroup arrangements, informal training procedures, broadly defined jobs, and absence of specialist and technical functions.

These findings both supported and provided some explanation for earlier work (Burns and Stalker, 1961), which had identified that in turbulent situations (e.g. those faced by smaller companies, growing in new markets with new products) organic organisations developed, whereas more mechanistic, i.e. more 'structured', organisations were associated with more predictable and certain situations (e.g. larger, maturer companies, providing more stable products for more stable markets).

## *The information processing model*

The 'information processing' approach to the study of organisations (e.g. Thompson, 1967; Galbraith, 1973) came later. The relationship with conclusion 4 above will be evident, and it will be worth summarising some of this work, although it was not aimed specifically at the investigation of technology factors.

Galbraith argued that the degree of uncertainty facing an organisation influences the organisational structure that is chosen or evolves. As uncertainty increases, the amount of information to be processed in making decisions within the organisation also increases, and hence the organisation must take suitable steps in order that it might cope. Four organisational design strategies were identified, which, if employed, might:

1. Reduce the need for information processing by:
   (a) the creation of slack (i.e. spare) resources, and/or
   (b) the creation of self-contained tasks.
2. Increase the ability of the organisation to process information by:
   (a) the introduction of vertical information systems, and/or
   (b) the creation of lateral relations.

The creation of slack resources reduces the number of exceptions that occur simply by means of reducing the required level of performance. The creation of self-contained tasks implies the establishment of groups with all the resources necessary to perform the task, such groups affording increased flexibility, responsiveness, etc. The use of vertical information systems provides a supplementary mechanism for the processing of information and therefore avoids the overloading of conventional hierarchical communication channels. The creation of lateral relations enables the level of decision-making to be moved down the organisation and closer to the point at which the information originates.

Thompson, in attempting to develop theories for organisational design, pursued a similar approach. He too was concerned with the influence of uncertainties on companies. He classified companies in a manner that emphasised interdependence and co-ordination. For example, the degree of interdependence was seen to be high in mass production, where each stage of the process is *sequentially* dependent on a predecessor. In contrast, in a bank there is a form of *pooled* dependence, of each branch on a central office, but without direct interdependence of branches. In seeking to cope with increased uncertainty, Thompson suggested that sequentially dependent companies tended to introduce more planning and scheduling, and to pursue vertical integration. Pooled-dependence companies tended towards standardisation.

### *Coping with complexity and uncertainity*

This glimpse of some classic studies has highlighted the importance of uncertainty and complexity, and shown the need for the structure of organisations to be contingent on such influences. The 'task technology' studies showed that uncertainties evident at the level of tasks, workflows, etc., may be associated with certain organisational characteristics. The 'information processing' studies showed that uncertainties and complexities at the level of the firm might be associated with certain organisational responses. Any factors that give rise to increases in complexities and uncertainty faced by an organisation may prompt, or necessitate, some organisational response, i.e. some action designed to cope with the stimulus. Clearly changes in the external environment, e.g. market/demand changes, may be the cause of these effects for a company. The 'information processing' investigations and some of the 'post-Woodward' studies dealt specifically with this aspect. Technology may also be a cause of increased uncertainty or complexity, especially internal to the organisation and either at the company levels or at some lower level.

Notice that a relationship is seen to exist not only in the dynamic, i.e. change, situation, but also in the steady state. Changes in complexity and uncertainty, whether internally or externally originated, prompt changes in the organisation. A given level of complexity and uncertainty may be associated with given organisational characteristics. In both cases an appropriate 'match' must be the aim.

## Organisations for New Technologies

We can now use the model that we have outlined, and look more closely at the impact of (new) technology. There are two scenarios available to us:

1. Technology as a source of complexity and uncertainty.
2. Technology as a moderator of, or a means to reduce or contain, complexity and uncertainty.

## *Technological complexity and uncertainty*

Most of the studies that we have cited view new technology as a source of increased complexity and uncertainty for the organisation. There are, in effect, two stages to be considered.

1. *Technological change*, i.e. the change period, during which something new is planned and introduced, e.g. the period associated with the introduction of new plant or processes, and/or the introduction of new products or services that have major new technological ingredients.
2. *The post-change state*, i.e. the period following 1, when the changes have been implemented and absorbed and something approaching steady state has been reached.

These two stages may be identified for any particular example, but in fact, because of the rate of change of technology, many companies may in practice be affected most of the time by 1. While the introduction of new process technology in a department may not be followed immediately by its replacement by newer technology, the steady-state periods are reducing, and for most larger companies change situations will be evident, in some part, at most times.

By definition, change is a source of complexity and uncertainty, so the above argument adds weight to the hypothesis that new technology is a major source of such impacts on organisations. However, even in the post-change/steady technology state, the newly introduced technologies may increase complexity and uncertainty. For example, the unreliability and unpredictability of systems have been found to have a major impact on companies and personnel. In one survey of new technology in manufacturing (Wild, 1986), the need for manufacturing managers continually to monitor the performance of 'automated' equipment, to anticipate and rectify failures, and to compensate for the inadequencies of the equipment were major tasks and concerns. Further, the nature of the new technology may, inherently, contribute to complexity and uncertainty. For example, in manufacturing, the trend is to introduce flexible manufacturing systems capable of processing a large range of different types of product, as a mixed flow, in very small batch sizes, through the same set of facilities. As yet fully automated, plant-wide FMSs are not common. Typically FMS 'islands' are established that must then interface with other parts of the manufacturing system. Further, there remains the need for support facilities, e.g. handling, maintenance, quality control, etc. Since an FMS provides mixed flow with little short-term repetition, uncertainty is increased for the other and for the support parts of the system. This compares to the relatively predictable repetitive situation characterised by large batch production, or single model processing, which the FMS may have replaced.

So, there are adequate grounds to support the hypothesis of technology as a source of complexity and uncertainty, and it will be appropriate to review the

**Table 2** Some organisational responses

Increased complexity and uncertainty, e.g. resulting from the introduction of new technology, may be associated with:

Greater delegation, and participation in decision-making.

Less emphasis on formal role definitions, procedures, rules and practices.

Less emphasis on formal decision-making structures, and reduced concentration/centralisation of authority.

Reduced emphasis on sophisticated control procedures.

Less 'mechanistic' management organisation. Greater emphasis on 'organic'-type organisations.

Less emphasis on formal, detailed job descriptions.

Greater use of 'matrix'-type organisations.

Greater use of project and team working.

Greater use of semi-autonomous workgroups.

Less emphasis on specialist or technical functions.

Greater mobility of individuals between jobs.

Less division of labour and specialisation.

Greater use of 'temporary' organisations.

Greater use of self-contained tasks, or task groups.

Greater use of interdependent activities, and reliance on informal co-operation.

organisational consequences. Table 2 summarises the main organisational characteristics or responses identified in the previous section, and the discussion below picks out some of the principal organisational options available to management seeking to tailor the organisational structure.

## Matrix organisations

A typical 'matrix' type organisation structure is illustrated in Figure 3. there are two dimensions—outputs or projects, and inputs or functions. Individual staff are associated with both dimensions, e.g. they may be associated with the computing function while also working with a project team. The outputs or projects dimension represents the work demands on the organisation. The inputs or functions dimension is the resources. Such organisation structures provide flexibility (individuals can serve multiple projects and/or move between them, as required), but also allow individuals to retain membership of functional or 'common expertise' groups. Communication and decision-making may be facilitated and implementation improved. Tasks (project) orientation is emphasised, and the arrangement provides for the existence of temporary,

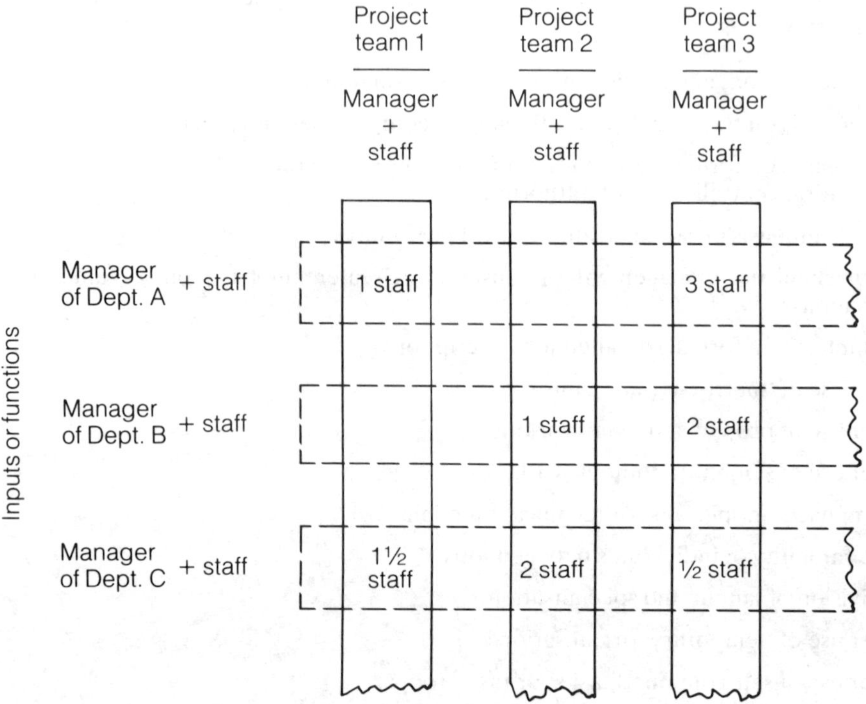


**Figure 3** Simple matrix organisation structure.

self-contained, possibly largely autonomous project groups or teams. Matrix organisations may be appropriate in dynamic situations, but care must be taken to prevent conflict of interests and roles, and excessive use of meetings. They may prove costly to operate. Matrix organisation is a means for providing Galbraith's 'lateral links', and represents a type of contingency approach to organisation design (Dawson, 1986).

## Group working

The creation of semi-autonomous, self-contained work teams or groups is consistent with many of the responses listed in Table 2. Such teams may represent one dimension of a matrix organisation or have more permanence. In general such a group will have been established for a positive purpose, have defined boundaries of responsibility, and be self-regulating. The members of the group will possess most of the skills and knowledge relevant to their tasks, and have access to all relevant resources and such additional skills as may occasionally be required. Some of the design principles for the formation of such groups are illustrated in Figure 4.

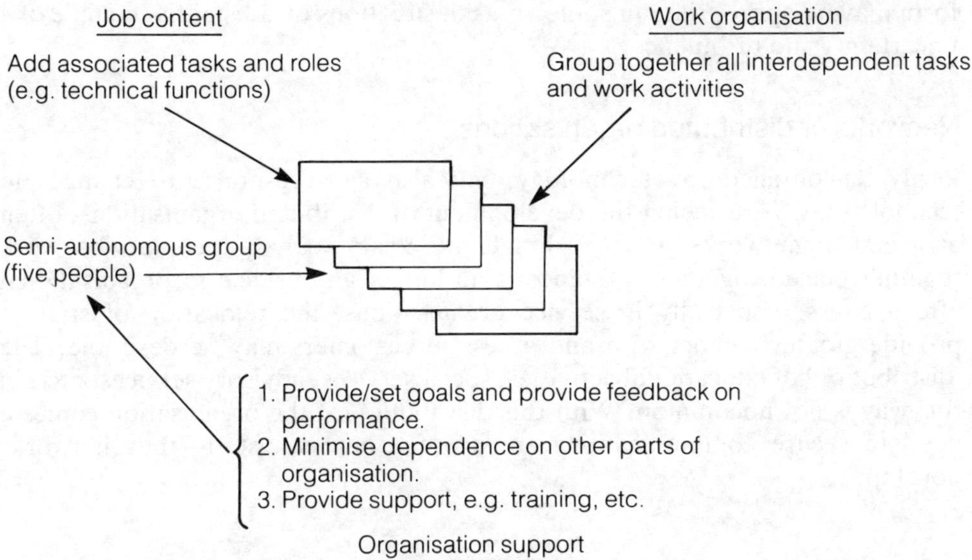


**Figure 4** Principles for group working organisation.

## 'Virtual' groups

Recently Taylor (1988) has defined a type of workgroup organisation that is similar to, but identifiably different from, the above. Taylor's 'virtual' groups are compared to semi-autonomous workgroups in Table 3.

Virtual groups are more informal: they develop, operate and dissolve. Individuals come together for a purpose(s) they identify. They work together for as long as is necessary, and then move on to form other groups. In fact at any time individuals may be associated with several 'virtual' groups, the extent of their involvement and contribution varying in a manner influenced by group needs and individual capabilities. Further, 'virtual' groups may have multiple (usually) related purposes or 'missions'. To a large extent 'virtual' groups represent an extreme example of the adoption of the responses identified in Table 2. They offer total flexibility and need little substantial hierarchical structure. They exist, and are required, in social, cultural, creative, entrepreneurial, academic, research and sporting activities. Such an organisational

**Table 3** Semi-autonomous versus 'virtual' workgroups

| Semi-autonomous workgroup | 'Virtual' workgroup |
| --- | --- |
| Formal | Informal |
| Work within fixed/defined boundaries | Work across boundaries |
| Defined purpose | Multiple purposes |
| Self-regulating | Self-organising |

form may be appropriate in some areas in situations of substantial complexity, uncertainty and turbulence.

### Networks or distributed organisations

Partly 'encouraged' by technology, but also as a response to changes in technology, we are seeing the development of distributed organisations, often arranged in networks. It is no longer necessary to have people who work together placed together. Distance is no longer an obstacle to organisational effectiveness. Especially in service organisations, the relocation of staff to provide greater contact with and access to customers may be desirable. The 'distribution' of educational activities, social services, advisory services, etc., in this way is not uncommon. With this distribution of the organisation comes a need to secure communications and contact between staff—thus networks develop.

## *Supportive and facilitating technology*

We have concentrated on technology as a causal or contributory factor in organisational change. While the bulk of the evidence available supports this view, the effects of the introduction of new technology may be neutral or even positive in this respect. New technology may have no relevance or may act as a moderator of, or a means to reduce, contain or eliminate uncertainty and complexity from other causes. In such situations the changes covered in Table 2 and those described above will not be pursued for technological reasons—but may be required as a response to external factors. New technology may also facilitate some of these changes. The development of technology for the support of management and decision-making activities (see Part 3) is potentially an example of this. Computer simulation, spreadsheets, decision support systems, etc., are supportive and may therefore facilitate delegated or distributed decision-making. New communications technology will facilitate distributed or networked organisations.

## *Evolving organisations*

Together with all of the above, i.e. the organisational implications of the effects of technology on complexity and uncertainty, technology will also influence organisations through its impact on employment and manning. In general, new technology will change job requirements in companies—if not the overall number of jobs, then certainly the number of jobs in particular areas. Process automation, especially, will affect employment requirements—for example, fewer direct process jobs will be needed. Manual work will be replaced by technology and some routine planning and control activities and routine communications activities will be eliminated. In these circumstances managers find themselves responsible for fewer subordinates and dependent upon more peers. So, for this reason *also*, the shape or structure of the organisation changes.

We can hypothesise that, under the impact of process technology change, organisational structures evolve—moving from the traditional pyramid structure to something more diamond-like. The reduction in the number of 'direct' workers at the bottom of the pyramid narrows this part of the organisation. This effect continues as the extent of automation afforded by technology increases, until a complete level of the organisation is eliminated. Meanwhile, the process has already begun at the higher level, and so the effect continues up the organisation. In parallel with this effect is the greater dependence of the organisation on technical specialists and support staff. Often the number of such staff expands, and thus the higher levels of the organisation widen, creating the diamond shape to which I have referred. Over time this level of technical specialism also moves up the organisation so the diamond shape becomes shallower, with fewer levels in the organisation.

This might be seen as a general pattern of change that may be happening in parallel with the implementation of any changes of the type outlined earlier, i.e. the macro- and micro-level organisational changes associated with technological change.

## A Pragmatic View

In this chapter, I have attempted to deal with a complex issue. A large part of the complexity derives from the difficulty of adequately defining two terms. I have taken a simple approach. I have avoided extensive scrutiny of the numerous definitions. I have been pragmatic. However, it will be worthwhile to conclude with cautionary comments:

1. Much of the research that has been conducted in this area has focused on 'technology', using definitions that would not easily be recognised by present-day managers.
2. Many of the studies that have been underaken have pursued 'simple' relationships, e.g. by examining whole organisations.

In fact technology is a systems concept. It is not a simple dimension, but rather a cluster or matrix of things: materials, processes, software, practices, etc. The extent and manner of the usage of new technologies may be different in each part of the organisation. The rates of change may differ. Furthermore, each part of the organisation, i.e. companies, departments, groups, divisions, sections, areas, etc., may be run in a different manner, and may relate differently to technology change. So there can be no simple 'rule of thumb' to relate technology to organisation. For this reason, as managers or technologists we can only realistically hope to be aware of some of the things that might influence the relationship and some of the common features of these relationships, in order to be able adequately to deal with that part of the organisation in which we are involved or for which we have responsibility.

In this chapter I have confined the discussion to an overview. Some implications have been noted, and some directions for development identified. Now some of these issues can be taken up in the discussion of the roles of managers in the next chapter.

## References

Burns, T. and Stalker, G. M. (1961) *The Management of Innovation*. London: Tavistock.

Child, J. (1984) New technology and developments in management organisation. *OMEGA*, **12**, no. 3, 211–24.

Dawson, S. (1986) *Analysing Organizations*. Basingstoke: Macmillan.

Galbraith, J. (1973) *Designing Complex Organizations*. Reading, MA: Addison-Wesley.

Taylor, J. (1988) Paper presented to International Conference on Joint Design of Technology, Organization and People Growth. October, Venice.

Thompson, J. D. (1967) *Organizations in Action*. New York: McGraw-Hill.

Wild, R. (1986) Changing manufacturing technologies and policies, and the role of manufacturing managers. *International Journal of Operations and Production Management*, **6**, no. 4, 27–41.

Woodward, J. (1965) *Industrial Organization: Theory and Practice*. Oxford: Oxford University Press.

# 2.2 *Management Roles and Skills for New Technology*

*Adrian Campbell and Malcolm Warner*

## Interdependencies

It will be clear from the previous chapter that the innovations currently being implemented in industry and business involve both technological and organisational systems. The emphasis overall must be on *integration*, and so the technological and organisational aspects have themselves become increasingly interdependent. For example, in manufacturing, the introduction of computer-integrated manufacturing (CIM) must involve both the technological system and an organisational approach. Without the organisational 'effort' the full benefits are unlikely to be forthcoming. Implementation has to be accompanied (and preceded) by strategic reappraisal of organisational aims, structures, practices and attitudes (Campbell, 1989).

The era of the simple technological 'fix', whereby specific innovations could be applied successfully in isolation from each other, is now remote, if indeed it ever existed. In this chapter we shall develop some of the issues dealt with in Chapter 2.1. We shall refer again to organisational structures, but primarily in connection with management roles and skills.

## Priorities and Outcomes

The discussion in Chapter 2.1 raises several questions. For example, given technological change, will managers have *more* or *less* flexible roles? Will they have *more* or *less* distcretion in decision-making? Should management become *more* participative or *less*? The answers to such questions depend, in part, on *why* new technologies are being introduced. As Child (1984) has pointed out: 'Managers will normally have several goals in introducing new technology . . . the emphasis between these is likely to vary according to the priorities and purposes or their organisation and the context in which it operates.' To this we can add that different managerial groups are likely to have different goals regarding technology, which may be related to different perceptions of the organisational environment and what its implications are for strategy.

As we have seen, markets and the changes taking place within them provide one explanation for the extent of the interest in new technology: 'In order to remain competitive, many firms are investing in new technology . . . because markets are becoming more complex and possibly more differentiated [and] the environment in which firms operate has become more variable and uncertain. Even if the technological state-of-the-art is often ahead of the market, new manufacturing technologies are opening many marketing directors' eyes to new markets' (Warner, 1986, p. 280).

Market factors provide a link in the causal chain, leading to responses involving new technology (although, as stated above and in Chapter 2.1, the process can work the other way round). The changes in markets and technologies in turn imply a variety of organisational effects, and therefore effects on managers. For example, in recent years markets for manufactured items have become fragmented, with more differentiated markets, smaller batches and greater customisation (Campbell and Warner, 1988a).

While in practice it is still likely that the 'giant company' will continue to provide the context within which the activities of smaller firms operate (Pratten, 1986), the increased 'hiving off and sub-contracting to smaller firms represents one of the major organisational changes taking place in the manufacturing sector' (Shutt and Whittington, 1987).

There are thus a number of discernible trends towards decentralisation, reintegration of skills, smaller batch production and increased customisation (linked to higher expectations from customers regarding quality). So, in many situations, and not only in manufacturing, the use of new, potentially more flexible technologies makes economic, strategic and technical sense. If organisations operated according to classical free-market theory, whereby firms are guided by 'market forces' to make appropriate decisions, there would be no organisational problem. However, as Senker (1988) has pointed out, the actions of management and of 'market forces' are not the same thing. Management always have the discretion whether or not to innovate in a particular fashion, and the consequences of their not doing so, or doing so badly, may *not* immediately become apparent. This choice is a significant factor even where new technology is introduced as part of an integrated strategic approach. The potential for success may still be constrained by managements' inabilities. Thus before looking at the implications of new technology for management, it is important to emphasise that management attitudes are crucial in terms of whether (or how) such changes take place. As Long (1987, p. 257) argues: 'A major challenge for many managers will be to modify or even abandon their traditional classical approaches to management. The new technology can be used to help make the organisation a dynamic, adaptive entity, or it can be used to fight a rearguard action against such adaptation.'

## Impact of New Technologies on Management Roles

Management structures may emphasise either the vertical or the horizontal. If the former, roles and tasks are sharply delineated and communication (in theory) is rooted in the lines of hierarchical authority. If the latter is emphasised there is more delegation of responsibility, the flow of information is more open, and there is a blurring of organisational barriers. Organisations of the latter type tend to be more adaptive, as we saw in Chapter 2.1, and it is such organisations that are more likely to be successful in periods of rapid change. At present organisations are being influenced in the direction of a more horizontally oriented structure, not only by the pace of change generally, but also by the intrinsic qualities of the technologies involved. This process is likely to be accelerated (and complicated) by the expanded networks of informal links that occur where task groups and project or matrix structures are involved (Campbell and Warner, 1988*b*).

These forms of 'team work' are likely to become an enduring feature of organisation where new technology is concerned. The above trend confirms the findings of Rothwell (1984) on new technology and the elimination of the supervisory level. If such elimination occurs, it would be very likely to amplify the role played by middle management, along with the latter's need, in present circumstances, to liaise more closely with customers and suppliers regarding design, deliveries, quality and service. A different scenario is of course possible: the potential provided by technology for closer monitoring by senior management could be used by them as a protective cloak under which they would have few reservations about delegating more responsibility to supervisory level. Middle management, therefore, deprived of both its authority-enforcing and information-passing roles, could be rendered increasingly redundant.

Such a projection may depend on how the term 'middle management' is defined. If, for example, decentralisation of functional or business units is increased, and this decentralisation takes place on account of senior management's increased ability to monitor performance from a remote point, are the managers who head up such satellite units 'senior managers', or does their increased day-to-day accountability to the organisational 'core' of senior managers mean that their status is in reality lower than it would have been considered in the past?

## Impact on Management Skills

What new skills will 'technology managers' need? First, they will need to be more technically aware, if not technically qualified. German firms have an existing advantage as a greater number of their managers have technical or engineering degrees. Japanese managers, too, have technical qualifications of a high order (see Handy *et al.*, 1987). Second, they need specific management training to acquire the appropriate 'interpersonal skills' needed for operating within more complex, less hierarchical organisational structures (in addition to the likelihood of more external contacts). The nature of managerial work, and therefore the skill requirements also, will change, from first-line supervisor to top management, according to Boddy and Buchanan (1986, p. 181). Managers need to be able to manage the process of change, in which communication plays a very significant part.

With the introduction of information technology into increasingly complex areas of the business, the management of change becomes an intrinsic part of the management role, rather than being associated with one-off projects. Whether in financial services, retailing or manufacturing, competitive pressure and expansion in the range of viable applications mean that there is frequently no clear break between successive stages of a particular innovation project, and no clear dividing line between projects. This will have a number of consequences for management roles, which will be acted out within what may be termed a 'culture of permanent implementation'.

## Technological Awareness and Organisational Politics

There may be an increase in the power of whichever internal function has most control over, or awareness of, systems development and application. This is important in terms of the relationships between senior and middle management. Although senior management reserve the power to make strategic decisions, those lower down, if they have an awareness of the strategic dimensions of systems-related decisions, may exercise a hidden influence. This potential will clearly be pronounced to the extent that senior management lack such an awareness (Campbell *et al.*, 1988). The lack of such an awareness at higher levels may, however, suffocate strategically appropriate use of new technology, or the adjustment of strategy to derive the full benefits available from its use. A survey of consultants' views on systems implementation (Campbell, 1989) revealed this lack of awareness to be a recurrent problem in terms of effective choice and implementation of systems. Senior managers in many firms were said to fail to comprehend the scope of changes implied by a move into or up-grading of information technology. As a result, consultants found themselves forced to play down some of these implications in the early stages for fear of losing the contract. Where internal managers rather than external consultants play the leading role in initiating change of this type, a similar sequence of events is not uncommon. With internal managers there may, however, be some benefit in political terms, since the same lack of awareness at higher levels that necessitates the use of guile on their part may consolidate their 'ownership' (see Pettigrew, 1973) of the technology concerned.

The studies referred to in the previous section found decisions on the acquisition, implementation and up-dating of technology to be highly political in terms of the managerial interactions involved. This is in apparent contrast to the findings of a larger survey by Hickson *et al.* (1986). The latter study found decisions on new technology (in most cases) to be 'not unduly complex or political', unlike other areas of decision-making, where the politicality derived from the range of functions and personnel involved (as in major reorganisation, for example), such that the decision became a 'vortex' drawing everything in behind it.

The reason for the discrepancy between these two views on technology is of fundamental importance for the subject of this chapter. In the study of Hickson *et al.* decisions were identifiably of political importance because they were *seen* to imply major changes for a wide range of different groups and for the functioning of the organisation as a whole. With new technology, according to our findings, the politics of decision-making were more hidden, and derived precisely from the fact that such decisions were *not seen* by senior managers to have far-reaching implications for the business, while others (whether inside or outside the organisation) did see such implications if any benefit was to be derived. This lack of awareness is more serious to the extent that each successive decision on technology may narrow the effective scope for manoeuvre on further decisions (Clark *et al.*, 1988).

## Managerial Misconceptions of Technology

The preceding section points to the need for a more thorough understanding by senior management of new technology and its strategic potential and implications.

Two fallacies are held by the more traditional managers: first that technological systems are ultimately 'pieces of kit to do a job' and unproblematic in terms of effective incorporation and utilisation; and second, that investment in such 'kit' is a one-off decision. In fact, it may mean the beginning of a complex series of projects involving modification, up-dating and recommissioning, to say nothing of the need to review and restructure organisational procedures and train and retain staff. Systems applications rely in the long term on the accumulation of knowledge within the organisation, not merely technical knowledge of data-processing, etc., but a broader understanding of the benefits systems can provide, what configuration of systems functions best suits the needs of the business and how best to implement with a view to achieving those benefits.

Matching technology to the needs of the business is not the only problem, however. Too often managers have doubts on what their strategic requirements actually are, and are therefore all the more likely to pursue the 'technological fix', hoping that expenditure on technology will of itself fill in the gap left by their lack of a coherent or forward-looking strategy.

Although overall strategy may be the preserve of senior management, such strategies only work through the development and implementation of sub-strategies at lower levels (Clark *et al.*, 1988, p. 33). The successful development and implementation of strategy relies on co-operation and communication at all levels. The findings of Rothwell (1985, p. 375) are not encouraging in this regard, showing decisions on technology to be usually 'top-down' in character, to the extent that supervisors and end-users are often the last to know about the nature of the changes proposed.

## Information Technology and the Delegation of Authority

Information technology provides the potential for three types of change in the way authority and initiative are handled within an organisation. First, the information available for strategic (or any other) decisions is greatly increased through computerisation, although it would be unwise to expect more information to produce better decisions automatically. The claims made for decision support systems, whereby managerial work is seen to have been 'automated', are as misleading as those regarding secretarial work, and result from the system designers having grossly over-simplified notions of how such work is actually carried out (Long, 1987).

Second, and more important, the development and communication of background data, objectives and performance indicators is greatly facilitated. This can result in more intrusive use of authority by senior managers, but at the same time it can also provide for more informed debate on and questioning of decisions (where this is possible), and (more commonly) the possibility of more strategically oriented behaviour by management at lower levels, who, through increased availability of data, are more aware of the parameters they are working in, where the key bottlenecks are, which areas would be the most profitable to concentrate on, etc. An example from recent research (Campbell, 1989) is that of a storekeeper in a large electronics company who, following the introduction of a computer networking system, spends the greater part of his time in discussions with customers and suppliers with whom he had no previous contact. This more 'entrepreneurial' approach at lower levels is one that has recently begun to be approved of and encouraged by a variety of large companies in different sectors. Providing the opportunity, in terms of both the technology and the organisational climate, does not of itself guarantee results. Companies have often neglected planning and related skills at lower levels, so that they are entirely reliant on individual flair when given the opportunities for a more cerebral approach than traditional fire-fighting and 'muddling through'.

Third, there is the communication involved with the development and application of information technology itself. Just as suppliers may have erroneous notions of what organisational roles comprise, so within the organisation, those responsible for introducing and implementing systems may be unaware of the requirements of the end-users within the organisation. The ability of all parties involved to be able to communicate effectively in such circumstances is, along with the awareness that such communication is necessary in the first place, central to the success of projects involving new technology. Similarly, management needs to be able to communicate its needs to the suppliers of the equipment and the consultants, if the latter are used.

This is particularly important where smaller firms likely to be lacking in-house expertise are concerned.

## The Role of Management Education in Technological Change

Management education should be focused more on the principles of technology and strategy, in addition to the principles associated with technology management in the more routine sense. Implementation, increasingly recognised as an area of difficulty, will also need to be addressed by specific courses.

Management training will be required to present change as an organisational feature, rather than as an occasional activity carried out within organisations. Students need to be encouraged in the direction of investigating the accelerated independence associated with new technological systems, rather than the more traditional approach geared to the management of discrete projects. Recent technological developments and the resulting need for a more 'holistic' approach to change have also led to demands for different approaches to accounting and project evaluation techniques, and management training in this area would require more emphasis on recent alternatives such as 'throughput accounting' (Waldron, 1988), which have been developed with this broader approach to technology in mind.

In addition to the communication skills mentioned above, there is also a need for changes in traditional management attitudes, which may obstruct change. Adaptive behaviour may be developed and encouraged through appraisal systems and in-house courses. We do not of course suggest that change is not problematic. It may frequently be the case that sweeping changes are put into effect on someone's initiative high up in the organisation, on the basis of half-digested principles or straightforward self-interest. Those forcing the change may even employ the rhetoric of the 'adaptative and flexible organisation', while presenting an unadaptive and inflexible face to opponents of any part of the strategy, who are deemed to be 'resistant to change'. Such initiatives are misconceived, not least in that they regard change strategies as exclusively top-down in nature. The point is that the more that attitudes open to change are encouraged, the less one particular person or group's view of change is likely to dominate unconditionally or undeservedly. Scenarios of the type referred to immediately above may well result from a prior period of stagnation.

## The Up-Grading of the Training and Development Function

The challenges facing companies in terms of the need for more flexible managerial skills necessitate a change in the importance and role traditionally allotted to the training and development function, a shift that may already be seen occurring in a variety of larger businesses. With the decline of union influence, the traditional personnel function has also declined in influence, although commentators such as Long (1987, p.174) see in information technology generally an opportunity for personnel. However, personnel professionals have failed to take adequate advantage of this. In Britain, our own findings have found many instances of a more strategic role being played (discreetly) by revitalised training functions within larger organisations (Campbell and Warner, 1988*c*). As awareness increases of skills and adaptiveness as a factor in gaining competitive advantage, the strategic scope and influence of the training function may well increase markedly. Training and similar 'cost-centres' (business development, etc.) may provide some of the momentum for change in organisations where the line management is either weak or conservative, or where there is a growing body of opinion that the hierarchically oriented approach is not proving either proactive or responsive enough.

## Summary and Conclusions

### *The changing context of management*

Figure 1 draws together the different causes and effects at work in terms of technology and management skills and roles. First, changes in the external environment of the business, which we have summarised as market

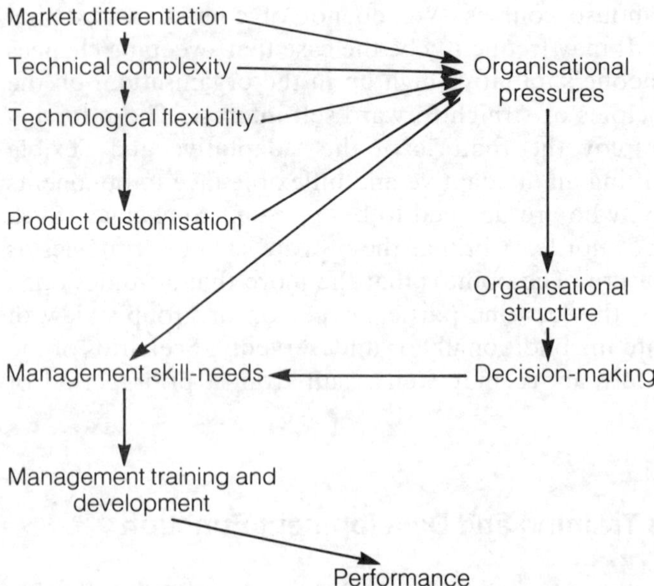


**Figure 1** The relationships between environmental, technological and organisational changes.

differentiation but which also incorporate the geographical expansion of previously stable markets and tighter competition on both quality and price, place greater demands on technological capability. This leads to an acceleration in the development of technologies that are more complex in design and flexible in application, facilitating differentiation and customisation of products and services, and implying an emphasis on satisfying specific demands at the same time as optimising the efficient allocation of organisational resources. Changes in technologies and markets in turn place pressures on organisations to innovate, both in terms of their overall strategy and in terms of their use of technology. These adaptations, if they are to be successful, require changes in both organisational structure (in many cases making it less hierarchical or functionally oriented) and the criteria adopted in decision-making on new technology, which, as they become a more consistent focus of management attention, need to be seen more in terms of the overall needs of the business and less in terms of piecemeal cost-saving.

## The management education agenda

These changes require the development of managerial skills in several areas:

1. The ability to arrive at a more coherent strategy for the business in line with external trends (for senior managers).
2. The ability to assess the contribution to that strategy that may (or may not) be made by different technologies (for middle and senior managers).
3. The ability to translate the advantages of new technologies into proposals that may be successfully justified within the framework of the management's overall strategy (for technical specialists, engineers and managers in favour of specific technological changes).
4. The ability to operate within the new terms of reference and investment criteria brought about through advances in information technology (for senior managers).
5. The ability to operate effectively without the structure provided by functional hierarchies and traditional lines of communication (for management at all levels).
6. The ability to delegate authority to a more autonomous and highly skilled workforce (for supervisors and middle managers).
7. The ability to use information technology as an aid in day-to-day planning and development of each area of the business (for middle and junior managers).

These skill requirements demand a significant increase in *education* as opposed to training. Training in terms of how new technology may be used on a day-to-day basis is also required, increasingly for managers as well as for other employees. By 'education' is meant a shift in managers' perceptions of the role of information technology in the business and the potential to use it to alter their own roles in the organisation. For example, if information networking between functions and between firms is to produce benefits, there has to be some dissolution of organisational boundaries and a greater fluidity in managerial job descriptions (as in the example of the storekeeper cited earlier).

## *The learning organisation*

This development of initiative and exploratory activity at all levels is a characteristic of what Hayes *et al.* (1988) term 'the learning organisation'. In their view, too many managers look for the optimal solution, or believe they have found it, rather than occupying themselves with continual improvement. This continual learning rather than a particular strategic configuration is seen as the key to long-term success. This is not an approach that is recognised and rewarded in many organisational career systems, which tend to revolve around clear-cut indicators of performance, and which may reflect over-hurried change or crisis management salvaging of the status quo. This again means changing attitudes at the highest levels. Senior management must be educated first if useful learning is to occur on any scale at operational level or below.

To conclude, there is a need for far greater resources to be directed towards management education and development (as well as workforce training and education) if the investments currently being made are to produce benefits across the organisation as a whole, and this should be mirrored, in institutions concerned with management education, by a greater emphasis on the problems and potential of new technology at strategic and operational levels. More fundamentally perhaps, organisations should be oriented towards continual 'learning' itself as an indicator of performance, even though, as with most information technology applications, the benefits of learning are not amenable to quantification.

## References

Boddy, A. and Buchanan, D. A. (1986) *Managing New Technology*. Oxford: Blackwell.

Campbell, A. (1989) Innovation and organisation: a consultancy perspective. IDOM (Innovation, Design and Operations Management), Aston Business School (research report).

Campbell, A., Currie, W. and Warner, M. (1988) Innovation, skills and training in Britain and West Germany. In Hirst, P. and Zeitlin, J. (eds) *Reversing Industrial Decline*. London: Berg Press.

Campbell, A. and Warner, M. (1988*a*) Workplace relations, skills-training and technological change at plant-level. *Relations Industrielles*, **43**, no. 1, 115–30.

Campbell, A. and Warner, M. (1988*b*) Organisation for new forms of manufacturing operations. In Wild, R. (ed.), *International Handbook of Production and Operations Management*. London: Cassell.

Campbell, A. and Warner, M. (1988*c*) Strategic choice, organisational change and training policies: case studies in high technology firms. Management Studies Research Paper, no. 8/88. University Engineering Department, Cambridge.

Child, J. (1984) New technology and developments in management organization. *Omega*, **12**, no. 3, 211–24.

Clark, J., McLoughlin, I., Rose, H. and King, R. (1988) *The Process of Technological Change: New Technology and Social Choice in the Workplaces*. Cambridge: Cambridge University Press.

Handy, C., Gow, I., Gordon, C., Randlesome, C. and Moloney, M. (1987) *The Making of Managers*. London: Manpower Services Commission, National Economic Development Council and British Institute of Management.

Hayes, R. H., Wheelwright, S. C. and Clark, K. B. (1988) *Dynamic Manufacturing: Creating the Learning Organization*. New York: Free Press.

Hickson, D. J., Butler, R. J., Cray, D., Mallory, G. R. and Wilson, D. C. (1986) *Top Decisions: Strategic Decision-making in Organizations*. Oxford: Blackwell.

Long, R. (1987) *New Office Information Technology: Human and Managerial Implications*. London: Croom Helm.

Pettigrew, A. (1973) *The Politics of Organizational Decision-making*. London: Tavistock.

Pratten, C. (1986) The importance of giant companies. *Lloyds Bank Review*, January.

Rothwell, S. G. (1984) Supervisors and new technology. *Employment Gazette*, January, 21–5.

Rothwell, S. G. (1985) Supervisors and new technology. In Rhodes, E. and Wield, D. (eds), *Implementing New Technologies: Choice, Decision and Change in Manufacturing*. Oxford: Blackwell, 374–83.

Senker, P. (1988) International competition, technical change and training. Science Policy Research Unit and Imperial College Papers in Science, Technology and Public Policy, no. 17.

Shutt, J. and Whittington, R. (1987) Fragmentation strategies and the rise of small units: cases from the North-West. *Regional Studies*, **21**, no. 1, 13–23.

Waldron, D. (1988) Accounting for CIM: the new yardsticks. *Industrial Computing*, supplement, 36–7.

Warner, M. (1986). Human-resources implications of new technology. *Human Systems Management*, **6**, 279–87.

# PART 3

# *The Technologies of Management*

We have considered the management of technology, and the impact of technology on management roles. The management of technology is a problem area, and a responsibility of management. The development of technology in the organisation affects that organisation's structure and the role and responsibilities of those managers operating within it. Thus the first and second part of the book have dealt, in effect, with managerial problem areas associated with technological development. Here, we consider the ways in which new technology might assist the manager.

The chapters in this part of the book deal with the technologies of management, i.e. the use of technology in the managerial process. The first chapter provides some background information in the development of those technologies now available to the manager, together with an overall view of topics dealt with in greater depth in the three other chapters. The second, third and fourth chapters deal with the type of technology now available to the manager, 'on his or her desk'. In addition, there are 'pointers' to future developments, which are intended to assist the manager and to improve managerial productivity and effectiveness.

# 3.1 Technology for Managerial Productivity and Effectiveness

*Peter C. Bell*

Management must be aware of and responsive to technological innovation in products, in services and in production methods, but also, importantly, to the changing technology of management itself. The final section of this book is about new developments in management technologies: technologies aimed towards producing a leaner, more effective and more productive management. This chapter provides an introduction and overview. The remaining chapters in this section deal in greater depth with three of the topics introduced here.

Technology to aid managers is emerging from several directions. Many of these developments owe more to advances in computers and computing than they do to great conceptual leaps by researchers. In fact, many potentially important developments have lost credibility by being 'discovered' long before we have had the computer power available to permit their implementation. For example, as early as the middle 1960s, Stafford Beer (1985) was talking about a situation room where the senior management of an organisation could sit and watch the operation of the enterprise, and could investigate and implement decisions to control its direction. Unfortunately, the computer, information gathering and monitoring, and control technologies of the 1960s were nowhere near advanced enough to implement this idea in a major corporation. Beer's idea was, therefore, not taken very seriously at the time, but his concept has persisted; we now see situation rooms in several organisations that have fairly rigid management rules (for example, the military, engineering, e.g. telecommunications or nuclear power plants, and space exploration). As our technology continues to develop, we can expect to see attempts to implement Beer's situation room within a firm.

Technologies that are directed towards improving managerial productivity and effectiveness are evolving from research in management information systems (MIS), management modelling that has its roots in management science and operational research (MS/OR) and artificial intelligence (AI). While these have historically been distinctly different research streams, today researchers in all three disciplines are very much concerned with developing improved methods to make use of data, models and knowledge to improve managerial efficiency and effectiveness. Developments in MIS, MS/OR and AI have been greatly influenced and stimulated by the evolution of computers and computing, which has provided increasingly powerful engines to implement new ideas.

## Developments in Computing

Data processing using punch cards and tabulating machines began with the work of Hollerith in the 1880s, but Hollerith's tabulating machines are not considered to be computers, since they had to be 'programmed' by complex setting of switches and wiring. In the late 1940s, mathematician John von Neumann developed a device that could store a program in the same way that it stored data, and the computer age began.

The period beginning in about 1950 and continuing to about 1970 was the age of the large computer—what we now call the mainframe. The emphasis during much of this period was on building bigger and faster machines, with software designed to be fast and efficient. Initially organisations would have only a single large mainframe, but as technology developed some decentralised data processors emerged, having several mainframes linked through telecommunications (primitive networks). The technological highlight of this period was possibly IBM's announcement, in 1964, of an entirely new line of computers called the System 360. This system, developed at an estimated cost of $5 billion (throughout, all dollars are US dollars), included many different processors, input, storage and output devices that were all compatible. These devices could be linked together to allow mainframes to be tailored to individual customers for the first time.

The outstanding feature of this period was the technological development that continually reduced the cost of large-scale transaction processing and data storage. This technological development continues in the supercomputers of today. These supercomputers are specialised machines designed to be the ultimate calculators. While their data input and output speeds are relatively modest, the speeds at which they can 'crunch' numbers make them valuable in computationally intensive applications (e.g. weather forecasting, military weaponry, advanced scientific uses).

During the 1970s, the movement to decentralised computing and the arrival of communications between computers, together with the saturation of the market for large mainframes, prompted manufacturers' interest in smaller systems designed for smaller organisations. These minicomputers had less data storage space, operated at lower processing speeds and were physically much smaller than the mainframes. Since the minicomputer buyers had lower levels of computer skills, the manufacturers became less concerned with raw power and more concerned with helping the 'user'. The result was less powerful systems that were easier to use and more 'friendly' for the less experienced user. Recognition that, for many firms, the cost of software greatly exceeded the cost of hardware led to compatibility becoming an important marketing feature—why should organisations have to rewrite all their programs when their increasing demand for data processing necessitated a switch to a larger machine?

In 1969, the already high energy level of the computer industry received a quantum boost with the invention at Intel Corporation of the microprocessor—a computer on a single wafer of silicon. Initially personal computers (PCs) built around a microprocessor were marketed as kits to home electronics hobbyists until, in 1977, Apple began marketing the Apple II as an assembled unit. A major event in microcomputer history occurred in August 1981 when IBM Corporation announced its entry into personal computing with the IBM PC.

The announcement of a PC from IBM changed the nature of the personal computer business, even though the IBM PC was not a very technologically advanced machine. While the Apple II had been an astonishing market success, it was seen by business as a 'toy' suitable for home use rather than as a serious alternative to mainframe or minicomputer computing. The IBM PC carried the reputation of the major presence in business computing, and organisations bought IBM PCs to do serious business computing (although many early purchases of IBM PCs were cannibalised from IBM's other business machine markets, particularly electric typewriters and the DisplayWriter word processing machines). The growth in worldwide PC shipments has been astonishing: from about $3.8 billion in 1981 to more than $27 billion in 1987, with the number of PCs installed worldwide increasing from 200 000 in 1977 to more than 80 million in 1987.

In August 1984 IBM announced the personal computer AT, a machine based on a new generation of microprocessor—the Intel 80286. The AT was faster, by about a factor of 5, than the PC and offered increased main memory (RAM) and disk storage capacity. These developments hastened a major revolution in computing: in 1977, before the introduction of the first PCs, the total computer market of about $17 billion was split about three-quarters to mainframes and one-quarter to minicomputers; by 1983, dollar shipments of PCs had reached those of minicomputers, with PCs and minicomputers accounting for about half the hardware market; in 1987, worldwide dollar shipments of PCs almost equalled those of mainframes.

The traditional view of computer hardware is summarised in Table 1. This view has blurred significantly since 1985. We now have minicomputers (sometimes called superminicomputers, e.g. the IBM 4381 series) that are more powerful than many mainframes, and desktop computers that are more powerful than many minicomputers. The term professional workstation is used to describe these powerful (and expensive, often selling for more than $100 000) desktop computers. The professional workstation appeals to PC users who need greater computing power for scientific, engineering or artificial intelligence uses, or for heavily graphic applications such as computer-aided design. These workstations have generally been based on the Motorola 68xxx family of microprocessors, which are superior computationally but more expensive to implement than the Intel family (examples are SUN, XEROX, or MicroVAX).

In today's business market-place, the predominant PCs are manufactured

**Table 1** Traditional categories of computer hardware

| | Personal computer | Minicomputer | Mainframe | Supercomputer |
|---|---|---|---|---|
| Speed (millions of instructions per second) | <2 | 2–20 | <25 | <1200 |
| CPU storage (millions of bytes) | <1 | 1–8 | 2–16 | 32–2000 |
| Cost | \$500 to \$10 000 | \$10 000 to \$500 000 | \$250 000 to \$5 000 000 | \$2 000 000+ |
| Examples | IBM PC, AT, PS/2, Apple IIe, Macintosh | IBM System 34, 36, HP 3000, VAX 11/785 | IBM 200, 3033, CDC Cyber 170 | Cray X-MP, 2 |

by IBM or are IBM 'compatible' (i.e. will run most of the same programs as the IBM-made machines), although Apple retains a strong presence in one particular business sector—desktop publishing—with its Macintosh. Workstations have been becoming more popular (and less expensive), but the emergence of '386' PCs may limit the future of the workstation.

In September 1986, the microcomputer market-place was again excited by the announcement of the Compaq Deskpro 386—the first of a new family of PCs based on Intel's 80386 microprocessor. These new PCs, which were joined by the IBM PS/2 models 80 and 70 in 1987, provide much the same computing power as the workstation but in an 'IBM compatible'. The different capabilities of the three generations of PCs based around the Intel family of microprocessors can be seen in Table 2. It seems likely that 386 microcomputers will come to dominate the PC market-place quite quickly

**Table 2** A comparison of the three generations of IBM PCs

| Microprocessor | 8088 | 80286 | 80386 |
|---|---|---|---|
| IBM product | PC and XT | PC AT | PS/2 Models 70, 80 |
| Date of introduction | 1981 | 1984 | 1988 |
| Main memory (RAM) | ≤256 kB (PC)<br>≤640 kB (XT) | ≤640 kB | ≤16 MB |
| Disc drives | | | |
| Flexible | 180–360 kB | 360 kB–1.2 MB | 1.4 MB |
| Fixed | none (PC)<br>10 MB (XT) | 20–30 MB | 110–300 MB |
| Internal clock speed | 4.88 MHz | 8–10 MHz | 20–25 MHz |
| Approximate relative power | 1 | 5 | 10–15 |

because of the credibility that IBM Corporation has achieved as a supplier to this market, and the fact that entry level 386 PCs will be available at low cost very soon.

The PC has emerged as the preferred delivery device for systems for use by managers. This appears to be the result of a combination of a number of factors, particularly the greater user-friendliness of the PC versus a terminal connected to a mainframe or minicomputer, the emergence of highly graphic, easy-to-use PC software, and the appearance that the PC is a more private computing environment where the executive can experiment and make mistakes without a feeling of being watched. Advances in telecommunications now enable us to link executive PCs with a host of other equipment.

Telecommunications is now a vital part of hardware technology. The stand-alone mainframe or PC processing jobs in batches is history. We now think in terms of networks of computers (PCs and mainframes and minicomputers) sharing storage and printing devices, and capable of electronic messaging with each other and with other national and international networks.

The managerial workstation of the next few years is likely to be a networked 386 PC rather than a terminal to a mainframe. Connecting the PC to a network will provide the executive with access to corporate data, external databases and electronic communications. For the first time, we will have PC computing power on the executive's desk that is capable of running the kinds of systems that researchers have identified and developed to assist the manager. A persistent difficulty with this kind of thinking has been the 'software lag': the fact that advances in software have not kept up with the evolution of hardware.

Software is characterised as systems software and applications software. Systems software, which is generally purchased with the hardware, includes the operating system and the languages used to program the computer.

The operating system is the program that performs the essential operations needed to process users' jobs through the computer. Early mainframe operating systems were single-user, batch processing systems, but the move to multiple users with several jobs being processed simultaneously (multitasking) resulted in a significant increase in operating system complexity. The operating system for a modern mainframe is a very large and complex program designed to optimise the performance of the computer in the face of real-time demands from many interactive users at the same time as a heavy demand for batch processing.

The need for mainframe operating systems to provide flexibility for the advanced user results in their being seen as complex and unfriendly to the average user. Minicomputer manufacturers spent considerable effort on making their operating systems friendlier and easier to use, since they recognised the importance of ease of use in the smaller business market. This trend has continued with the PC, in part because the PC is a much simpler machine than either mainframe or minicomputer.

The first Apple PCs used the CP/M operating system, a rudimentary, single-user, batch-oriented system much like that of the early mainframes. The popularity of the IBM PC led to the MS-DOS operating system (called

PC-DOS when implemented on IBM PCs), which was specifically tailored to the instructions set of the Intel 8088 microprocessor, becoming a popular standard. The IBM AT used this same operating system, even though MS-DOS did not make full use of the much richer instruction set of its 80286 microprocessor. The first 386 PCs also used this same operating system, even though the 80386 microprocessor incorporates a whole new set of features into its design (e.g. virtual memory management and the ability to address 4 billion bytes of main memory).

In 1988, a new operating system known as OS/2 became available from the IBM Corporation and Microsoft Corporation. OS/2 is essentially an operating system for the 80286 microprocessor that makes use of some of the functions of the 80286 that are not in the 8088 instruction set (e.g. protected mode operation and multitasking). OS/2, however, has appeared nearly four years after the hardware it was designed for (the IBM AT and the 80286 compatibles). With the advent of the 80386 machines, our operating systems are now a complete hardware generation behind. There is, at the moment, no sign of an operating system designed for the 80386 machines. Indeed, history suggests that such a system is at least five years away—by which time we will have '80486', perhaps even '80586' microprocessors. As a consequence, much use is still being made of MS-DOS, and attempts have been made to update MS-DOS to accommodate some features of newer PCs.

MS-DOS was first designed when most PCs used less than 640 kilobytes (kB) of RAM and 640 kB was thought to be vast; sufficient for all possible PC applications. The 640 kB memory limit was built into PC-DOS and has been very difficult to alter. All versions of MS-DOS prior to version 4.0, which was released in the autumn of 1988, strictly limited users to 640 kB of memory. Consequently, in early 1989, software developers have only just begun to think in terms of PC software that requires megabytes of memory; we do not yet have a very good idea of what a 2–10 MB software package will look like or of its capabilities. One thing seems clear: a good deal of the extra available program size and of the development effort will go into making the programs easier to use.

Software developers and users who program their own applications make use of a computer language. Languages are generally recognised as having developed through four generations:

- *First generation* languages are machine-specific codes that are understood directly by the computer and are time-consuming to write.
- *Second generation* languages are assembly languages that use common instruction labels that can be translated into machine code by a computer program. Assembly language programs are not very portable, and can only be run on one specific family of computers.
- *Third generation* languages (3-GLs) freed programmers from specific hardware requirements; Fortran, COBOL, BASIC, etc., are produced in standard versions. Compilers can translate programs written in these languages into machine code to be executed on their equipment. The

3-GLs require the programmer to specify the steps that must be performed in order to carry out a task, and were the mainstay of data processing departments in developing business application systems for many years. The demand for new programs from end-users has swamped program developers and prompted the development of the 4-GLs.

- *Fourth generation* languages (4-GLs) emerged as a solution to the applications development backlog. The average corporate end-user now faces a wait of more than three years from the time that a new system is requested to the time that development begins. One response to this backlog has been the appearance of languages that are sufficiently easy to use for end-users to build their own systems.

The distinctive feature of a 4-GL is that the user specifies what must be done and the software is able to determine the steps required to do it. Fourth generation languages also facilitate fast program development (although program run times are often slow), tend to use natural language types of statements and operate interactively. The major types of 4-GLs are data query and retrieval languages (e.g. SQL), report generators (e.g. NOMAD), and application generators (e.g. FOCUS and MAPPER). As an example of their efficiency, an application written in FOCUS has about one-tenth the number of program lines and can be completed in about one-fifth the time of an equivalent program in a 3-GL.

The popular PC spreadsheets (VisiCalc, Lotus 1-2-3, Excel, etc.) and the financial planning languages (e.g. IFPS) are interesting hybrids. These are end-user programming environments, and when used by the end-user for direct problem-solving are not considered to be languages. However, these tools can also be used to develop systems for use by others (for example, using macro command languages). When used to program in this way, the programmer must specify the procedure to be used to perform the calculations and, from this perspective, the spreadsheets are 3-GLs.

Applications software is programs designed to perform specific computational or data processing tasks. MIS researchers have long been interested in the different types of applications software and of the often critical role of applications software within the organisation.

## Management Information Systems

Our understanding of the role of computer-based information systems in the organisation has evolved over time. The earliest types of systems that were implemented were transaction processing systems, which primarily maintained records and automated routine clerical tasks (e.g. payroll systems, invoicing and billing systems or accounting record keeping). As these types of systems were improved, and new systems developed to provide managerial reports and statistical analyses, and to allow data queries, the more general name management information systems was applied. With the move towards the 'electronic office', office automation systems were implemented which maintained appointments, kept address lists and handled tasks such as word processing, data storage and retrieval, and printing address lists for bulk mailings.

A pivotal idea that has emerged from MIS is a type of information system known as a decision support system (DSS) (see Chapter 3.2). Keen and Scott Morton (1978) used the term 'decision support' to imply the use of computers to:

1. assist managers in their decision processes in semi-structured tasks,
2. support, rather than replace, managerial judgement,
3. improve the effectiveness of decision-making rather than its efficiency.

This idea caused a considerable controversy within both MIS and MS/OR. A variety of MIS professionals attempted to improve on Keen and Scott Morton's definition of DSS, and many divergent views appeared. House (1983) summarised the situation.

> . . . no generally accepted definition of such systems (DSS) can be realistically stated at this point in time. The majority of current systems which are characterized as DSS do appear to be flexible, do deal principally with unstructured problems, and are at least partially interactive.

Management scientists and operational researchers who had been developing models to help decision-makers address all manner of problem situations and who, in many cases, had worked closely with managers in reaching decisions for complex, 'unstructured', problems were surprised at the fuss the term DSS has caused. Naylor (1982) expressed a typical view of DSS from MS/OR: 'DSS is a redundant term currently being used to describe a subset of management science that predates the DSS movement.'

With the benefit of hindsight, it is clear that MIS professionals were not the first people to use DSS, but their labelling 'decision support system' and their discussion of the formalisation of DSS has been pivotal for, I suggest, one simple reason: the DSS idea proved to be marketable. Managers had proved reluctant to buy systems designed to tell them what they must do, but were willing, perhaps even eager, to buy systems that were advertised as supporting them and helping improve their effectiveness. As Naylor agonised in 1982, 'it seems that virtually every computer hardware and software firm in the industry refers to its products as DSS', MS/OR professionals and software developers have not been shy to recognise the marketability of the label DSS, and a host of products are now advertised under this label.

The degree of 'structure' within a management task is central to the DSS concept. Gorry and Scott Morton (1971) argued that the role of information systems should be considered in relationship to Anthony's (1965) three levels of management activity; operational control, management control and strategic planning. Operational control tasks are the lowest level of management activities and include those activities required to manage the existing operations of the firm; for example, securities trading, parts ordering, software

**Table 3** Gorry and Scott Morton framework with examples

| | Level of management activity | | |
|---|---|---|---|
| Degree of structure | Operational control | Management control | Strategic planning |
| Structured | Order materials | Derive master schedule | Lease or buy building |
| Semi-structured | Schedule equipment maintenance | Budgeting | Mergers or acquisitions |
| Unstructured | Hire operators | Hire management | Select a board of directors |

purchasing, etc. Management control tasks are usually performed by middle-level management and include those activities necessary to manage the management of the organisation; for example, hiring a new manager, setting budgets, setting production levels, etc. Strategic planning is undertaken by the most senior management of the organisation, and consists of those activities that change the nature of the organisation itself; for example, building a new plant, launching a major new product, buying out a competitor, etc.

Gorry and Scott Morton's framework identifies DSS according to the level of management at which decision-making takes place and the degree of structure in the decision-making process begin supported. Their framework with some examples is illustrated in Table 3.

A useful complement to the Gorry–Scott Morton framework is Alter's (1980) classification of the types of assistance provided by DSS:

1. Retrieving a single item of information.
2. Providing a mechanism for *ad hoc* data analysis.
3. Providing pre-specified aggregation of data in the form of reports.
4. Estimating the consequences of proposed decisions.
5. Proposing decisions.
6. Making decisions.

The more structured the problem, the higher the level of assistance that can be provided. DSS can make decisions for structured tasks at the operational control level where workable solutions can often be derived using modelling techniques, and can usually be implemented without great opposition from management since the cost of mistakes is often quite low. There are a large number of such systems in place; for example, IBM's IMPACT system manages inventories of a large number of items, extrapolating usage data to forecast demand, computing reorder quantities for items that are nearly out of stock and printing orders to suppliers for restocking. Management's role in such a system evolves from one of deriving solutions and implementing them to one of monitoring the system and identifying exceptions and abnormalities.

At the other end of the scale, many unstructured strategic problems are very difficult to automate much above the provision of raw data. Here management feel that they have a unique skill and, further, the cost of poor decisions is very high. Thus it is difficult to develop a DSS that might be useful, and even harder to persuade management that an information system could provide useful help with their decision process.

Between these two extremes, there exist a host of situations where there is sufficient structure to develop some form of useful DSS and where there is a willingness on the part of management to accept this kind of support if it improves managerial effectiveness—that is, if DSS use leads to improved decision-making. The challenge facing DSS developers and advocates is to push the area of application of DSS towards less structured problems at higher levels of the firm.

The obvious high pay-off to be obtained from DSS for very senior management to use in strategic activities has led to the appearance of two types of senior management information system: executive information systems for individual managers, and the electronic boardroom for groups of decision-makers.

An executive information system (EIS) is a computer-based tool for personalised electronic information delivery to the senior manager. EIS are designed to be used by senior, non-technical, executives with little training, using few instructions, and generally provide vivid colourful output. The EIS would typically reside on the managerial workstation of the non-technical executive who can use the EIS to access, create, package and deliver information on demand. The present state-of-the-art EIS (reflected by systems such as Commander (Comshare Inc.)) are extremely easy to use, highly graphic, data query and review tools. A typical EIS would include the ability to select raw data from a database according to different perspectives (i.e. by product or by division) at varying levels of detail, to view routine exception reports, to view electronic documents (personalised to the receiver), to access stock prices, industry news and news services, and to produce graphics for presentations. Some EIS also include blank areas that can be filled in by in-house analysts or modellers to provide a powerful vehicle to add value to data for senior management decision-making.

The electronic boardroom serves the same functions as the EIS, but is designed for use by a group of senior executives. This usually requires facilities for projection of the computer screen output, perhaps more than one screen simultaneously, and facilities for multiple access to the EIS, including, perhaps, by a technical specialist in the boardroom who can respond to management requests in real time. One variant of the electronic boardroom is the decision support room.

The decision support room or situation room has an engineering analogue in the control room at, say, a nuclear power plant. The concept of a room where the senior executives of an organisation can sit together and run the organisation was first developed by Stafford Beer (1985), who devised such a room for the government of Chile to run the Chilean economy. The situation room differs from the electronic boardroom in that it has to include facilities both to sense the state of the organisation and to implement managerial directives in real time. Situation rooms are a common feature of military command and mission control for space agencies but, for the moment, their use by management is limited to occasions when a decision-making group of senior managers must temporarily wrestle through some difficult, but important, decisions.

## Management Models

The development and use of mathematical or symbolic models to aid management decision-making began during the Second World War when a group of scientists was assembled at RAF Fighter Command to provide a scientific point of view on various operational problems. This group first worked on the problem of translating data on approaching hostile aircraft received from the air defence warning system into an organised fighter response but, during May 1940, they were involved in some very high-level and critical decision-making. As Larnder (1979) recounts:

> When the Germans opened their offensive against France and the Low Countries, Fighter Command was quickly involved to the extent of ten of its Home Defence squadrons. These had to be maintained and operated from airfields on the Continent. [By May 1940] the British losses were running at a rate of some three squadrons every two days . . . On 14 May, [Air Chief Marshal Sir Hugh] Dowding learned that the French Premier was asking for an additional ten squadrons and that Churchill, because of his strong sense of loyalty to Britain's ally, was determined to accede to the request. On the morning of May 15, [Dowding] invited Larnder to see him and . . . finished by saying, 'Is there anything you scientists can suggest bearing on this matter?'. Anything that could be done had to be done before he would leave for the Cabinet meeting two hours later.
>
> So, at the suggestion of E. C. Williams, a rapid study was carried out based on current daily losses and replacement rates, to show how much more rapid this would become if the losses were to be doubled while the replacement rate remained constant. For ease of presentation, Larnder converted Williams' findings from numerical to graph form . . .
>
> [At the Cabinet meeting] Dowding, feeling that he was making little headway in dissuading the Prime Minister from his determination to reinforce France, got up, walked round the table and said to Churchill, 'If the present rate of wastage continues for another fortnight, we shall not have a single Hurricane left in France *or* in this country'. . . He laid his graphs in front of the Prime Minister. In Dowding's considered view, 'That did the trick'. Not only were the requested ten additional squadrons not sent to France, but of those already there all but three were returned to the United Kingdom within a matter of days.

This was clearly a pivotal decision. The Battle of Britain, which began about two months later, was fought with an average strength of 650–700 fighter aircraft: with 20 squadrons of fighters in France for a single week, the loss of 250–260 aircraft was expected.

This incident from the early history of management modelling exemplifies three characteristics of good management modelling. First, the decision-maker (the Prime Minister) was very high up in the organisation; second, the model, presumably based on a fairly straightforward exponential decay curve, was produced in timely fashion and was appropriate to the problem; and third, some effort was made to present the results of the modelling in a way that the decision-maker could understand.

The mathematical or symbolic model, solved by hand, was the mainstay of management modelling before the arrival of the computer. Often, solving the model was a time-consuming and laborious task involving rooms full of people with calculators working long hours (often overnight). The new electronic computers were first seen as particularly good calculators to solve this computational problem (their accuracy was a real benefit), but it was soon recognised that the computer could drive other kinds of models, particularly simulations.

The emerging importance of simulation models of systems prompted interest in simulation modelling software (GSP in Britain, GPSS and SIMSCRIPT in North America) in the early 1960s. Up to this point, the computer was very much a batch processing device: the modeller produced a program, which was submitted to the computer, and the output was returned some time after the job was submitted.

The appearance of interactivity has had a profound impact. At the managerial level, interactivity has led to managers routinely using the kind of programming-free modelling environments offered by spreadsheet programs (such as Lotus 1-2-3). For the model builder, we now have sophisticated modelling environments that permit the interactive construction of highly interactive models that make extensive use of colour and graphics to aid communication between user and model.

Our understanding of the appropriate modelling support for a particular management task has evolved alongside the DSS movement. Optimising models, such as mathematical or dynamic programming (which provide the solution having the maximum profit or minimum cost) have been widely applied to highly structured operational control problems (e.g. the day-to-day operation of an oil refinery). It is now recognised that optimising algorithms can be incorporated into DSS to allow the user to address less structured problems that may be formulated for optimisation in many different ways. Use of such a DSS enables the user to judge the sensitivity of the solution arrived at to various possible formulations of the problem.

Another broad class of models is the analytical models (e.g. queueing theory, Markov processes), which do not provide an optimum solution but allow the calculation of performance statistics for a given set of input parameters (for example, queueing models provide mean customer waiting times for a wide variety of different types of queueing systems). Results for these kinds of models were once published in tabular form, but now, routinely, these models are used in interactive environments that allow the user to vary the inputs and observe the resulting changes in the performance statistics.

Simulation modelling has become a very big business (see Chapter 3.4). Computer simulation is now routinely used to review the operation of manufacturing plants before construction begins, to design warehousing and materials handling systems, in military operations and gaming, and in a broad range of service sector applications (including airports, banks, etc.). The technique of simulation modelling has moved away from batch simulation towards visual interactive simulation, where the decision-maker can watch a dynamic, colour graphic display of the output of the simulation model, and can interrupt the model, change parameters or decision rules and continue.

The first visual interactive simulation models emerged when modellers ran into the problem of displaying the status of the model so that the user could see when to interact (and the effect of the interaction). After experimenting with the display of intermediate numerical results and transferring these results to physical models of the system, the modellers began producing dynamic computer-generated graphic displays. As our technology has developed, the quality of these displays has improved enormously: we now have simulation software that can produce multicoloured, three-dimensional, dynamic graphics showing the operation of large systems, and all on a microcomputer. Visual interactive simulation models, which have many of the characteristics of DSS, have motivated a more general modelling analogue of the DSS: the visual interactive model (VIM).

The term VIM originally applied to visual interactive simulation models, but has now been generalised to describe any system designed to support decision-making that includes computer-generated graphic displays of model status, run time interaction with the model and a symbolic or mathematical model or algorithm (Bell, 1985, reviews VIMs). There are many examples of the successful use of VIMs in organisations, particularly new plant or warehouse design using simulation, transportation planning or routeing using interactive computer-generated maps, and project management using computerised critical path tools.

Management modellers have also developed a second analogue of the DSS: the management model embedded in a data and model management system. A complex management model, say a large scale mathematical programming problem, might require the input of more than 50 000 variable values, and could well produce as many output values. When such a model is used frequently, it is worthwhile to construct an interactive management system around the model. Such management systems use sophisticated interface design ideas to allow interactive access, review and updating of input data, to prompt execution of the model, and to edit the model's output to produce informative displays. The difference between this kind of system and a DSS is one of philosophy; the management model 'solves' a structured problem and outputs the solution for implementation, while the DSS is a tool for a decision-maker to use to explore possible solutions to semi-structured or unstructured problems.

Modellers are paying increasing attention to the design of their model interfaces. Colour and graphics are now widely used to communicate model results to the user. The announcement of the NeXT computer, which includes facilities for high-quality stereophonic sound, suggests that, in addition to colour and graphics, modellers will soon be looking to incorporate a sound track into their models.

## Artificial Intelligence

Artificial intelligence (AI) has proved a difficult concept to define, but it seems clear that the subject that is now called AI emerged in the 1950s in attempts to get computers to think like humans (see Chapter 3.3). Originally, it was thought that if only a big enough computer could be assembled, then this machine would be able to replicate human thinking and learning processes. This research failed as we began to understand more about both computers and human brains, but several interesting developments have followed.

Beginning in the 1960s, AI researchers have focused considerable attention on heuristic search techniques. Many problem-solving processes involve some type of systematic search of possible solutions; for example, computerised speech recognition involves producing an electronic imprint of a spoken sound and then searching through a library of imprints to try to find a match. The game of chess attracted considerable attention during this period, since there exist a finite (but large) number of possible moves, and the 'best' move can (theoretically) be found by a systematic search. The problem is that there are so many possible moves that not all can be investigated in real time: some kind of rule-of-thumb (or heuristic) must be found to limit the number of moves that need to be evaluated. Modern chess-playing programs use very sophisticated searching rules and can beat all but the very best human chess players.

One development emerging from progress in search techniques is natural language processing. Natural language processing does not usually imply voice recognition, but rather involves transforming instructions written in something close to conversational language into machine understandable code. This usually involves breaking the sentence down into phrases or words and then searching lists of machine instructions to try to match machine instructions to the natural language instructions. If this were possible, machine translation of one language to another might also be possible, as well as communication between a non-computer literate senior manager and a computer system. Unfortunately, it has proved much more difficult than many thought; English is not a very precise language and there are many opportunities for ambiguity. At the moment most of the advanced development of natural language interpreters is in the area of front-ends to databases (examples are systems such as Intellect and Paradox), and in adventure strategy games.

Knowledge has also attracted considerable attention from the AI community. Two development areas are knowledge representation and knowledge-based or expert systems (ES). Knowledge has some apparent similarities with data, suggesting that organisations might develop and store 'knowledge bases' within the organisation, from which they could read and write knowledge. In practice, it has proved difficult to do this. Most corporate knowledge exists in the minds of employees; it is difficult to extract and store. A number of interesting issues exist; for example, who owns knowledge? In this milieu, AI has made progress in developing schemes to represent knowledge. 'Knowledge engineers' can 'debrief' individuals and store their knowledge in a way that others can access. Knowledge acquired from somewhere can be 'taught' new things and can be updated.

Knowledge-based or expert systems (ES) emerged from universities in the mid-1960s from efforts to build software to mimic human expert problem solving. General Problem Solver (Lindsay *et al.*, 1980), was an early attempt at a program to identify the steps needed to move from a starting position to a goal—this involves identifying 'operators' that can transform particular situations into other situations, and the application of a series of 'rules' that determine which operators to try to use at each step along the way. DENDRAL (Newall and Simon, 1963), a program that inferred the molecular structure of a chemical compound from mass and nuclear magnetic resonance spectral data, was built at Stanford University beginning in the mid-1960s and was the first specific ES. MYCIN (Shortliffe, 1976), built in the 1970s to diagnose bacterial infections of the blood, used some 500 'if then' rules to store its knowledge, a form of knowledge base that has become common.

There are many examples of ES (such as DENDRAL and MYCIN) that do not address business problems, but many examples of profitable stand-alone business ES now exist. One that is frequently cited is the XCON ES at Digital Equipment Corporation (DEC). XCON was developed by DEC to configure its VAX minicomputers. The system can produce a parts list from a customer order (many of which include unique system requirements) in a fraction of the time required by a higly skilled technical editor. XCON, which contains several thousand rules, has been used to configure all US and European orders for VAX systems since 1985, saving DEC an estimated $15 million annually. These savings come from increased throughput without increasing the number of technical editors, improved customer service through fewer shipping errors and better fit between the parts shipped and the customer needs (i.e. not shipping a 15 foot cable when a 10 foot cable is adequate).

Another commercial ES, the CATS system at General Electric Company of America (GE), illustrates some changes in management thinking that ES have caused. For 40 years, David Smith, GE's best diesel–electric locomotive troubleshooter, had advised locomotive engineers on the appropriate repairs to make at service shops around the United States. As Smith approached retirement, the traditional approach would have been to assign a young engineer as an apprentice to Smith in an effort to absorb Smith's knowledge before it left GE on his retirement. One problem with this approach is that the apprentice might leave GE, taking the knowledge away. Instead, beginning in 1980, GE decided to try to build a locomotive troubleshooting ES based on Smith's knowledge. Three years later, GE had the system (called CATS) installed on a PC in every GE locomotive repair shop. CATS enables a novice engineer to uncover the fault, and leads the engineer through the required repair procedures, including parts drawings and specific demonstrations of how the repair should be made.

Stand-alone ES, like XCON and CATS, may well prove profitable, but these types of sytems will not have the managerial impact that will occur when ideas from ES begin to be integrated into the major information systems of the organisation.

## Technology and Management

The role of technology in management can be considered in relationship to Anthony's (1965) types of tasks that managers perform within an organisation: operational control, management control and strategic planning. In performing these management tasks, managers make use of three important inputs: data, models and knowledge.

Data reach the manager from a wide variety of different sources. In some cases, the data consist of 'hard' facts and figures that may be collected in the organisation's database, or in another organisation's database. Other data may be 'softer'—opinions of colleagues or estimates of future activity levels. The manager is constantly viewing, validating, selecting and transforming data, seeking to establish what useful information the data hold about what is happening in and around the organisation.

Viewing data alone is rarely very informative. The manager makes use of a wide variety of models in seeking to use data. These models are often informal rules of thumb, e.g. our cost-of-goods-sold is usually about 40% of sales revenues. Most organisations, however, have a host of formal models including, for example, financial models that link the various items in the financial statements, production models that calculate how many sub-assemblies must be made to produce a certain volume of finished units, marketing models that relate how much must be spent on advertising and promotion to yield a certain level of sales, and allocation models that determine the best way to spread the scarce resources of the organisation around the various products to be produced.

Managers do not accept data without thought, nor do they accept the output from models without question. Each manager brings a wealth of knowledge to the task of managing the firm. Knowledge can be very general, such as a code of ethics, or can be very specific, such as the fact that the last time we raised our prices, the competition followed suit. The manager gains general knowledge through education and specific knowledge through experience doing various management tasks repeatedly. The experienced manager, that is, one who has seen a wide variety of situations before, is a valuable organisation resource.

The roles of data, models and knowledge in providing decision-making support were originally thought of as quite separate. The earliest DSS (and also many of the executive information systems of today) emphasise the data aspect of managerial support. These systems are designed to provide the manager with the capability to peruse the database of the organisation and to select, summarise and display data that are pertinent to some management task. Any modelling or knowledge to be applied to the data had to be done either in the head of the manager or after extracting the data from the system.

Management models, on the other hand, have traditionally attempted to provide optimum or 'best' solutions. This is possible when a problem can be formulated in an unambiguous way and 'solved' for the best solution. There are a great number of operational control problems that can effectively be treated in this way (e.g. resource allocation, transportation). Optimising management control problems is more difficult, but is being done increasingly in some instances—for example, systems (like OPT) that derive an optimum master schedule for a manufacturer. The optimisation of strategic planning problems is much talked about and increasingly attempted; a good example is Keeney and Raiffa's (1976) multi-criteria analysis of the site for a new Mexico City airport. The modelling approach incorporates the selection and use of pertinent data in the modelling, but the use of knowledge is less formal. Both the decision-maker's and the modeller's knowledge affect the solution arrived at, but the process whereby this knowledge is used is not spelled out.

Early stand-alone expert systems rely on knowledge alone to reach a decision (although the ES is itself a form of model). These systems attempt to mimic the qualitative, intuitive problem analysis process undertaken by many experts. Early expert systems development tools permitted no modelling and provided no access to data that were not explicitly encoded in their knowledge bases. Some recent ES development tools (e.g. GURU) integrate a spreadsheet and colour graphics with an ES, allowing models and an ES to be combined with a colourful, graphic interface.

## Systems to Integrate Access to Data, Models and Knowledge

In 1982, Sprague and Carlson proposed a general structure for a DSS (shown in Figure 1) that included both a database and a model base. Sprague and Carlson's DSS integrates access to data with the use of appropriate models. Just as the data exist in a formal structure (a database), so the models exist in a similar structure (the model base). The role of the DSS is to select both appropriate data and an appropriate model to maximise the value added to the data in responding to an inquiry from the decision-maker. The knowledge needed to solve the problem resides outside the DSS with the decision-maker, but this knowledge is brought to bear on the problem by including the decision-maker in the problem-solving loop.

The dialogue generator is an important feature of Sprague and Carlson's DSS; management modellers refer to this as the user interface. The role of the dialogue generator (or user interface) in the process of problem resolution in situations where there is no clear optimum solution ('semi-structured' or 'unstructured' problems in the language of DSS) is to reduce the problem-solving cycle time.

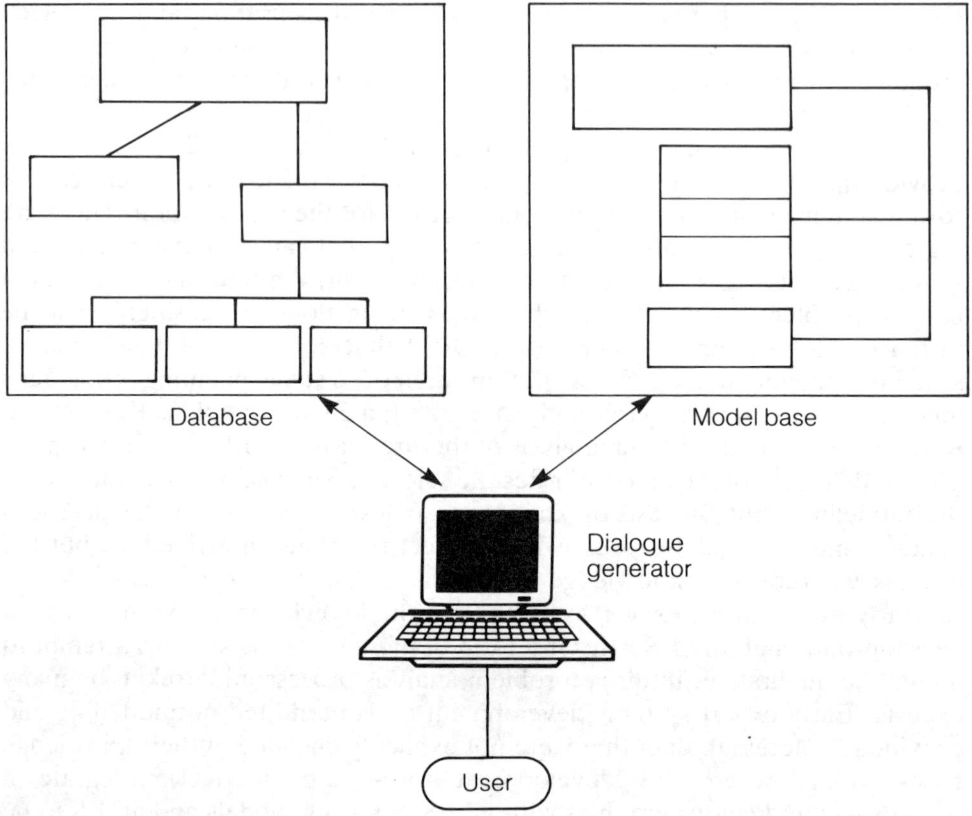


**Figure 1** Sprague and Carlson's general DSS.

Problem-solving proceeds via an iterative series of cycles: the decision-maker submits an enquiry to the DSS or model, which responds with some output. The decision-maker processes this information, incorporating personal knowledge, and submits a new inquiry. This iterative process continues until a solution is reached that the decision-maker judges to be acceptable, or the best possible. Figure 2 illustrates this process. Analysis of semi-structured or unstructured problems using a DSS requires the user to perform three functions on each iteration of the problem-solving cycle: first, to formulate an inquiry that can be understood by the system; second, to receive and comprehend the response from the system; and third, to analyse the response and think up an appropriate new enquiry. The cycle time for each iteration is the sum of the formulation time, the DSS or model processing time, the receiving and comprehending time and the user's thinking time.

The system processing time is an important design feature of a DSS. Although, occasionally, this can be quite long, in general, the more senior the manager, the faster the system must respond. Achieving fast DSS response times often requires clever data and model management, good (and computationally fast) models, considerable programming skill and, often, a powerful computer. The user's thinking time is less controllable, but can be reduced through training. Experience with the problem situation and experience using the DSS both contribute to reducing the time taken by the user to digest the DSS response and to formulate a new question.

The objective of the dialogue generator or user interface is to minimise the formulation time and the receiving and comprehending time. Increasingly, 'intelligence' is included in the interface; while this used to mean the use of a human 'chauffeur' to work between the executive and the system, increasingly ideas from expert systems and natural language processing are appearing in human–computer interfaces.

Designing interfaces for effective human–computer interaction involves elements of cognitive psychology, ergonomics (human factors engineering), software engineering and management. On the input side, the classical interaction styles (menus, form fill-in, command languages and direct manipulation, e.g. with a mouse or other cursor pointing device) each have advantages and disadvantages. On the output side, there is now a heavy emphasis on computer-generated graphics (including 'animation' or dynamic graphics), and on colour coding (in both graphic and tabular data), with windowing used to enable the user to combine elements of different displays on the same screen. It is useful to differentiate between two different forms of computer-generated graphics: iconic graphics and presentation graphics.

In an iconic graphic display, each picture element (or icon) maps to an element of the real situation—for example, a computerised street map, where the lines map to streets, the intersections to street corners, etc. A presentation graphic display presents a summary of some data—for example, the familiar bar charts, line graphs and pie charts. Computer-generated graphics appear to be most easily understood when they have some iconic structure that links them

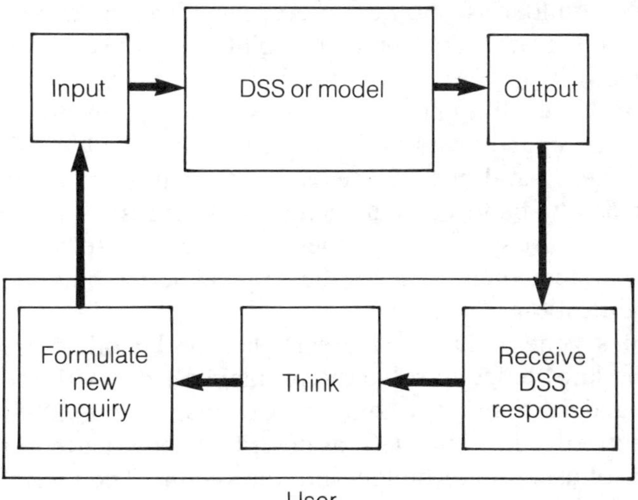


**Figure 2** Problem-solving with a DSS.

to the real-world problem—for example, interactive computer-generated maps have provided the basis for many successful DSS designed to support routeing decisions for vehicles or people around geographical areas. The WIMP (Windows, Icons, Mouse, Pull-down menus) interface, and the Lotus 1-2-3 interactive spreadsheet with 'pop-up' command menus are two widely admired and copied styles of DSS interface (for a more detailed treatment of interface design refer to Shneiderman (1987)).

## The Expert Decision Support System

There is currently a great deal of interest in integrating expert systems concepts with DSS to produce expert DSS (sometimes called expert support systems). There are a number of ways these ideas can be combined.

The expert front-end is an expert system that would sit between the DSS and the end-user, and would serve the purpose of a human 'chauffeur', interpreting user requests into DSS commands, and translating DSS output into information that the user could understand. An expert front-end could use its own knowledge to respond to DSS queries without always bothering the user, could provide intelligent advice to the user, could include a capability to recognise the user and to tailor the interface to different users, could add custom explanations of DSS actions and could act as a tutor. Expert models, which are expert systems included in the model base of the DSS, could be used to broaden the areas of application of the DSS, and to add more 'intelligent' and judgemental solution capabilities. Expert data management and expert model management would involve using an expert system to interpret user requests and to identify and extract the appropriate data or models to determine the DSS response. These types of systems could add judgement to model or data selection and, perhaps, provide maintenance and restructuring capabilities. An expert DSS builder is another interesting idea: an expert system that could play the role of the DSS builder, facilitating construction and/or modification of the DSS.

Of the expert DSS available, the most general is expert forecasting software. These packages (e.g. SmartForecasts, Forecast Pro) have colourful, graphic, user-friendly interfaces, and include several different forecasting models. The user enters the data to be forecast and a number of statistics about the series are computed. An expert system scans these statistics and selects a forecasting model, and the system then presents the original series plus the forecasts for review and modification.

In Sprague and Carlson's general DSS (Figure 1) data and models are included in the DSS, but the knowledge to solve the problem resides with the user and with the DSS builder, both of whom are external to the DSS. Attempts to incorporate knowledge into the DSS (as described above) attempt to internalise the knowledge of user and/or builder into the system. The logical extension of this idea would be an expert DSS where *all* the knowledge required to solve the problem was internal to the system. Such a system would be a decision-making system and, apart from human review and auditing, would provide the kinds of answers to semi-structured and unstructured problems that models are able to provide for structured problems.

## Decision-Making Systems

Developments in MIS, MS/OR and AI motivated by the need to improve managerial productivity and effectiveness appear to have the common objective of implementing managerial support systems that can speed up and improve managerial decision-making. Ideally, such systems could present a recommended decision for a broad class of problems within the manager's jurisdiction. We have these types of system now, but their use is restricted to structured, operational control problems. What is new is the progress that is being made in expanding the area of application towards less structured problems, and towards higher levels of the organisation.

This progress is being achieved through integration of MIS, MS/OR and AI ideas into systems that incorporate databases, model bases and knowledge bases, and that include intelligent data, model and knowledge base

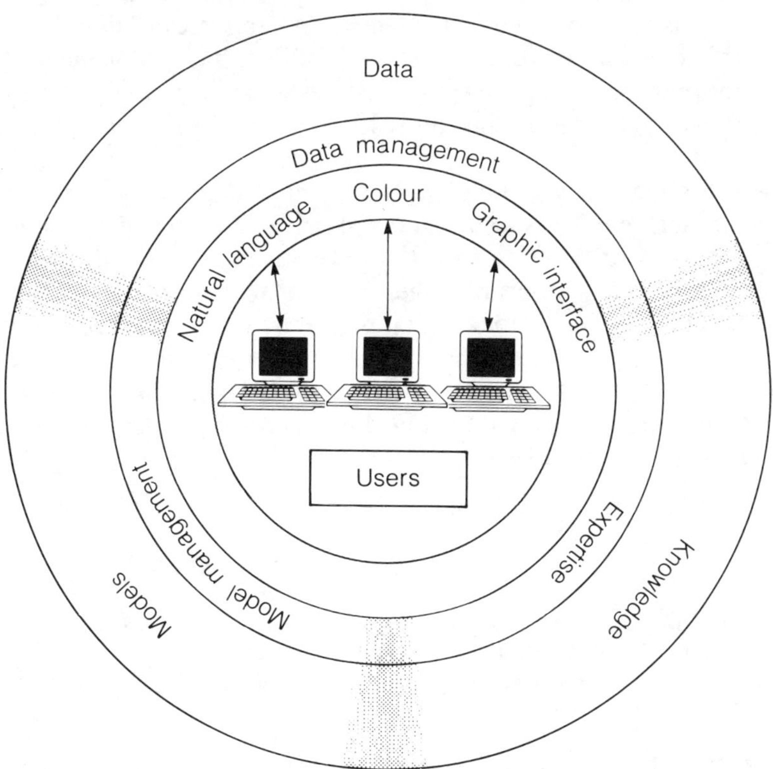


**Figure 3** Decision management system (from Bell, 1985/6).

management systems. A good deal of development is directed towards the design and construction of intelligent, graphic, natural language user interfaces that will enable one or more executives to use these complex systems to perform their day-to-day management tasks (Figure 3). Most of the emphasis to data has been on systems to support decision-making but, in some applications, it is only a small step to systems that both support and manage the decision-making process.

Decision management systems could cover a broad range of complexity. At one end of this range is a simple 'tickler' DSS that sounds an alarm when decisions are needed; for example, corporate financial managers often review their investment portfolios at certain times of the day and, following this review, go to the markets to make necessary revisions. A DSS could alarm the manager that it was time to make decisions and then prompt the user through the decision-making process acting as a DSS when needed. At the other end of the complexity scale, one can think of a decision management system that is the corporate analogue of a process control system, sensing changes in the state of the organisation, determining when these movements have taken key variables out of preset control limits, figuring out an appropriate (perhaps optimum) response and, with or without managerial review, taking steps to implement the response to bring things back under control.

Elementary decision management systems are here today; the challenge is to provide these systems for more senior managers. When we can build these types of systems for use at the board level, we will, at last, have implemented Stafford Beer's situation room of the 1960s. It will take some years to reach this level, but much is now going on that represents movement towards this objective.

Human managers have many unique skills that no computer system yet devised can duplicate, but managers also spend a great deal of their time doing work where computer support can improve both efficiency and effectiveness. Computer systems designed to support managers are the managerial equivalent of 'robots'; many expect that their impact will parallel the impact of shop floor robots on manufacturing employment. If so, organisations that are able to seize these technological opportunities will gain a competitive edge. The managers that remain will 'require unprecedented technical skills and the ability to relate these to their businesses' (Lowry, 1988).

## References

Alter, S. (1980) *Decision Support Systems: Current Practices and Continuing Challenges*. Wokingham: Addison-Wesley.

Anthony, C. (1965) *Planning and Control Systems: a Framework for Analysis*. Cambridge, MA: Harvard Business School.

Beer, S. (1985) *Designing the System*. Chichester: John Wiley and Sons.

Bell, P. C. (1985) Visual interactive modelling in operational research: successes and opportunities. *Journal of the Operational Research Society*, **36**, no. 11.

Bell, P. C. (1985/86) Emerging technology to improve managerial productivity. *Business Quarterly*, **50**, no. 4, Winter.

Gorry, A. and Scott Morton, M. S. (1971) A framework for management information systems. *Sloan Management Review*, **13**, Fall.

House, W. C., ed. (1983) *Decision Support Systems*. New York: Petrocelli.

Keen, P. G. W. and Scott Morton, M. S. (1978) *Decision Support Systems: an Organizational Perspective*. Wokingham: Addison-Wesley.

Keeney, R. L. and Raiffa, H. (1976) *Decisions with Multiple Objectives: Preferences and Value Tradeoffs*. Chichester: John Wiley and Sons.

Larnder, H. (1979) The origin of operational research. In Haley, K. B. (ed.), *Operational Research '78*. Amsterdam: North-Holland.

Lindsay, R. K., Buchanan, B. G., Feigenbaum, E. A. and Lederberg, J. (1980) *Applications of Artificial Intelligence for Chemical Inference: the DENDRAL Project*. New York: McGraw-Hill.

Lowry, P. (1988) Middle managers—a threatened species? *Daily Telegraph*, 27 October.

Naylor, T. H. (1982) Decision support systems or whatever happened to MIS? *Interfaces*, **12**, no. 4.

Newall, A. and Simon, H. A. (1963) GPS: a program that simulates human thought. In Feigenbaum and Feldman (eds.), *Computers and Thought*. New York: McGraw-Hill.

Shneiderman, B. (1987) *Designing the User Interface: Strategies for Effective Human–Computer Interaction*. Wokingham: Addison-Wesley.

Shortliffe, E. H. (1976) *Computer-Based Medical Consultation: MYCIN*. New York: Elsevier.

Sprague, R. H. and Carlson, E. D. (1982) *Building Effective Decision Support Systems*. Englewood Cliffs, NJ: Prentice-Hall.

# 3.2 *Decision Support Systems and Executive Information Systems*

*David Preedy*

## Introduction

This chapter examines the role of computer systems in improving the information available at the heart of today's major organisations. In particular it emphasises the different types of system available and how such systems can be combined to provide a co-ordinated solution to the main business requirements.

Clearly the ideal solution would be to provide a single unified management information system, which would support all the managers within the organisation, irrespective of their functional needs or their experience of the use of computers. With current technology, however, this turns out to be almost impossible to achieve. So this chapter starts off by describing a widely accepted categorisation of the specialist software that together meets the overall needs of the organisation's managers. It then examines in more detail each of the different tools, looking at their historical development and implemention requirments as well as the typical benefits that each can achieve.

## Components of a Managerial Information System

The starting point for any examination of the different types of management information system has to be to look at the underlying processes of management itself. For these purposes this chapter will concentrate on large organisations; in such environments it is necessary to establish a management reporting framework by which the performance of the different activities and their management can be assessed; the data about the individual transactions of the business provide an insufficient basis for management action and decisions. This means that a separate activity has to be established to generate management information, although clearly many of the data will originate from the operational systems of the organisation.

Once it is agreed that a managerial reporting procedure is needed, it is straightforward to recognise the three important roles needed to take effective action based on this information:

- *accounting*: this ensures that the information is collected in a timely and consistent manner in accordance with corporate standards and the requirements of any relevant external bodies.
- *analysis:* this involves taking the information provided by the accountants and interpreting it to identify and diagnose problems, to assess the impact of current policies and to estimate likely results of different courses of action.
- *executive decision-making:* at this level lies the responsibility for determining what actions should be taken based on the information available.

Depending on the size and management style of the organisation concerned, these roles may be combined to a greater or lesser extent. For instance, it is normal to expect board directors to undertake some limited analysis of the information provided in their regular management reports, but it would be unusual for them to become actively involved in exploration requiring the generation of different sets of data.

Each of these three activities offers a different perspective on the organisation's management information and has led to the development of a specialist type of software. The accountants, for instance, are primarily concerned with issues of data collection, integrity and robustness; they also require the ability to handle the complexities of currency conversion and consolidation, not just to meet the regular management requirements, but also to provide the data for the statutory reporting procedures. These are provided by consolidation and financial modelling packages, which are designed to enable detailed rules to be specified for routine execution. Analysts, on the other hand, tend to need more flexible access to corporate information, coupled with the ability to build and run simple models in short time scales. Their requirements lie at the heart of decision support systems (DSS). Finally we have the decision-makers—the directors and senior executives—who have to convert the information into actions and strategies. They need an effective means of assimilating information from a variety of sources and using such information as the focus of their management discussions. Recently the area of executive information systems (EIS) has been developed to meet such requirements.

Figure 1 summarises the three essential components of an effective management information system at the heart of a major enterprise. They cannot act effectively if they are isolated, either from one another or from the underlying operational systems, but each serves a distinct purpose and most large companies have recognised the need for a specialist tool in each category.

The borderlines between the three types of system are somewhat unclear and it is often feasible to use one software system to provide many of the

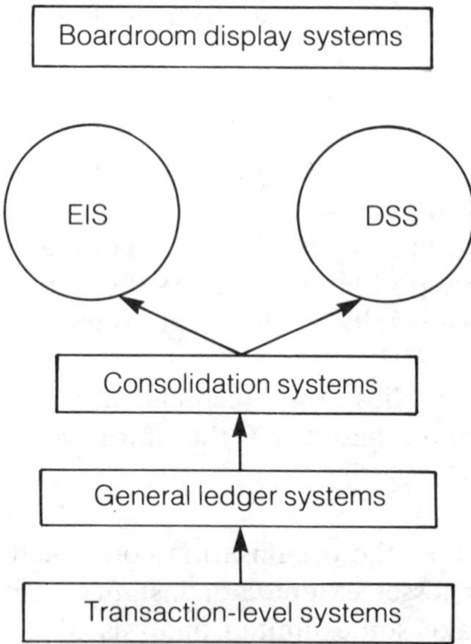


**Figure 1** Management information hierarchy (© Metapraxis, 1988).

facilities normally associated with a different area. Such solutions normally introduce some sacrifices of functionality or performance. For example, it is possible to use a microcomputer spreadsheet to create a consolidation system, but this may not handle the intricacies of currency conversion or inter-company eliminations as powerfully as a true financial modelling package.

## Consolidation and Financial Modelling

For the purpose of this chapter I am restricting attention to the process of management consolidation of accounts and am not considering the preparation of annual accounts on a statutory basis. This is not strictly management information since the process is primarily meeting external, rather than internal, requirements. The approaches described here may be useful to reduce the arithmetic chore of producing numbers for the statutory accounts, but other factors, such as system auditability and handling of journal entries, are also required.

### *Requirements*

The principal requirements of a consolidation system fall into three main areas.

#### Data input and validation

The consolidation system is the main area for input of corporate reporting data for the operating subsidiaries around the organisation. In some cases this input may come from underlying general ledger systems, in which case a suitable electronic transfer procedure can be implemented. Often, however, the results require adjustment by the local finance team before submission up the hierarchy and in these cases manual data entry is required. This means that suitable data validation routines are required to ensure data consistency, ideally at the point of input, but certainly before the data are released up the corporate hiearchy.

These requirements mean that a specialist consolidation system requires a comprehensive system for 'audit-trailing' so that the procedures of data input, editing and release are documented and controlled.

## Communications

The consolidation system is the main channel for transmission of data from one subsidiary up the management hierarchy. This requires effective communications, especially where the organisation operates in different countries, and also a controlled procedure for the release of information once it has been approved at subsidiary level. These requirements can either be met by a centralised approach based on a mainframe with access from remote locations, or by a distributed system, using local PCs for data input and validation, and autodial and answer modems for data transmission.

Historically, this communications requirement led to many consolidation systems being developed using time-sharing computer bureaux, which offered international network facilities. Such options are becoming increasingly available to companies with major centralised data-processing facilities, but even so the network has often had to be augmented with modem and telex input facilities to improve access from the further-flung parts of the organisation.

Many large organisations have diversified and operate in different industries or through fairly autonomous subsidiaries. In such cases the head office will not normally operate a large computing facility. Many of these companies developed the concept of distributed consolidation, where the individual reporting units operate a local PC-based data input, validation and reporting system. The results from the various subsidiaries are then transmitted via a communication link, using either autodial and answer modems or some temporary storage, such as an electronic mail service. At the head office (and any divisional or other intermediary consolidation points) there is a slightly more complex piece of software, which carries out the calculation aspects of the consolidation process.

## Calculations

The final aspect of a consolidation system is its capability to perform relatively complex arithmetic operations, and it is this capability that brings together the apparently different applications of consolidation and financial modelling. The common characteristics are that the 'rules' required are typically difficult to identify (at least for a systems analyst who is not a qualified accountant), complex and reasonably stable. This is because the rules for both types of system are determined primarily by the accounting and taxation requirements of the various countries concerned.

The consolidation process needs to carry out currency conversion. This may well involve: handling of multiple exchange rates for different countries and for budgets and plan data; the calculation of exchange rate effects, according to the organisation's accounting treatment; identification of non-monetary items, such as numbers employed and market share, which obviously should not be converted; different procedures to be applied to balance sheet lines for which year-to-date totals are meaningless; and so on. Likewise the aggregation of data from different operations must handle the elimination of inter-company sales and balances, and must avoid aggregating certain reporting statistics such as debtor days. However, these complications are balanced by the fact that once the rules have been defined, they can remain essentially unchanged for a long time.

Similar considerations apply to financial modelling, where the major benefit normally arises from seeing how the current rules for financial treatment apply to possible future scenarios. Here the complexities arise from factors such as defining tax rules with critical cut-off points, handling different tax regimes in different countries, and the iterative process of using profit predictions to derive balance sheet and cash flow figures, which in turn affect the interest payments that contribute to profit. Again the rules are complex but not rapidly changing.

For these reasons the normal consolidation systems may require a fair degree of expertise to program the calculation rules, and this programming tends to be carried out by specialists who are not involved in the monthly reporting process. With this background, it is normally advantageous to hold the calculation formulae separately from the data, with separate authority levels to make changes.

## Approaches to implementation

There are two popular approaches to consolidation and either may be appropriate to different types of organisation. The first is a centralised approach, which originated in time-sharing computer bureaus but has now been transferred in most instances to large in-house mainframe computers. This will be centred on a specialist software system, either specially written or based on one of the packages available; the most widely used of these are FCS[1] and System W[2]. The principle is that users enter and validate their data on the central system and the software ensures that data are released up the management hierarchy when released by the individual operating units.

The alternative is the distributed consolidation system. Under this arrangement each operating unit has its own input and validation system, normally based on a PC, and is responsible for transmission of its data to the centre according to an agreed timetable. While in theory the head office may not be concerned with the processing at the operating unit level, requiring only that the report is submitted to an agreed format, in practice most units will choose to use the same underlying software wherever possible. The software itself may either be a specialist consolidation tool, such as Micro Control[3], or can be a more general purpose tool, such as spreadsheet system. The purpose-built tools tend to have advantages when the calculations involved may be especially complex, or when the organisation requires a more formal control over the release of data.

In general the choice between a centralised and a distributed approach is likely to reflect above all the management attitudes towards autonomy and responsibility within the organisation. The centralised approach tends to enforce standards, for instance in the types of report made available to the management teams of each operating company. This can become a political issue if the local managers feel that the company standards are inappropriate for their own activity. On the other hand, the distributed approach encourages independence but its success depends on the co-operative efforts of the finance teams throughout the organisation. Moreoever, the distributed approach is much better suited to dynamic holding companies, which may be regularly acquiring and disposing of companies; it becomes much easier to integrate a new acquisition into the reporting system if the only hardware requirements are a personal computer and a modem.

Whichever overall approach is used, important decisions have to be taken concerning the software to be used. Essentially three choices are available: to use a general-purpose system, such as a spreadsheet package, and sacrifice some of the more complicated options for calculation rules and system management; to use one of the commercially available packages specially designed for consolidation and financial modelling; or to develop a bespoke system designed to meet the organisation's specific needs and computing environment.

There are several management issues that need be considered in making this choice:

- How much detail is required to be included in the monthly report pack. This will depend on the management style of the organisation and on the capability of the head office to use the data. A central team well equipped with an effective executive information system will be capable of interpreting a wider range of reporting lines than one relying on manual analysis. Many organisations find a compromise by reducing the standard pack and relying on exception reports to provide more detail when needed.
- The timeliness of the information. There is always a trade-off between the accuracy of the data and the time taken to provide them. Some companies recognise this factor explicitly by arranging for an estimated or 'flash' report as soon as the reporting period has finished. At the other extreme there are some elements of most management reports that are only fully reconciled when the annual audit is undertaken. Between these extremes,

there remains a dwindling number of organisations that feel that a quarterly management report is all that is necessary to maintain control of the business. These organisations tend to be very large companies that have decided that their divisions should be responsible for operational control of the businesses.

- Management style of the organisation. This will be reflected both in the detail required to be submitted each period and in the extent to which a centralised procedure is felt to be desirable.
- The nature of the business. A homogeneous organisation, such as a retail chain, will be able to define an agreed standard management information pack which can apply to each profit centre. This is likely to lead to a centralised approach to reporting based on a mainframe package fed electronically from underlying transaction level systems. On the other hand, a conglomerate may well have to report different performance indicators for different subsidiaries, albeit with a common core of key reporting lines. This environment is more likely to support a distributed approach. Likewise a vertically integrated company may have much more significant issues of inter-company trading than a conglomerate, the businesses of which do little inter-trading, all of which is likely to be carried out 'at arm's length'.
- The stability of the organisation. In a rapidly changing company, the technical and managerial constraints of imposing a centralised system may lead to significant delays in integrating a newly acquired subsidiary.

Likewise there are some technical issues that may influence the approach chosen. These include the following:

- Availability (or lack) of centralised computer systems.
- The diversity of computer systems in reporting units. Often the individual profit centres have different computing needs met by different, sometimes incompatible, computer systems. The distributed approach provides much easier access to a standard interface.
- Availability of programming support staff to develop and enhance the procedures.

## Benefits

In this section, I examine some of the benefits of computerising the consolidation process, and how to justify the investment. The major benefits derive from: the availability of the data for other parts of the overall management information system; reductions in the time taken to produce the routine management information packs; the release of the finance team's time to enable them to interpret the data being collated; and improved data reliability arising from better data validation on entry.

From this list it will be seen that many of the benefits are not derived directly from the consolidation system itself, but from the subsequent systems that use the data it generates. In this sense the consolidation system is a facilitator, without which other benefits from executive information and decision support systems are harder to achieve. The management information deriving directly from the consolidation system is normally provided in tabular form and on a month-by-month basis. As will be seen when we examine executive information systems, these two constraints reduce its effectiveness in the reporting process.

Financial modelling, on the other hand, should provide direct benefits by enabling the finance team to evaluate the financial implications of different scenarios. As with consolidation, the main object of the model is to generate the numbers for subsequent presentation and discussion elsewhere. The benefits are similar to those derived from decision support systems, which will be covered later. It is worth emphasising that in many cases the operational benefits of a strategic plan may be outweighed in strictly financial terms by the combined effects of issues such as currency movements, tax regimes, and so on; in such cases it is clearly essential that there is a reasonably accurate method to assess such effects and this can be provided by a financial model.

## Some case studies

### Centralised consolidation

One of the most comprehensive consolidation systems was developed internally by a major oil company in the mid-1970s. They decided to develop their own system because the tools then available did not give them the power to incorporate the detailed range of calculations that they needed. The main characteristics of their approach were the following:

- multiple multi-level hierarchies, so that they could model the legal as well as the management structure of the company
- local data input and validation, ensuring that only consistent data could be released up the corporate hierarchy
- reconversion of historic data at current exchange rates, so that trends could be examined without the distortions due to currency fluctuations
- comprehensive elimination of inter-company sales and balances at all levels of the organisation structure; this was especially important to cater for the widespread inter-company trading and financing undertaken by the company
- a wide range of standard reports that could be applied to any consolidation point on the company hierarchy.
- access to several years' back data so that trends could be compared.

As with all successful consolidation systems, the implementation has remained stable for many years, having been designed to allow for the normal management changes that may be needed. However, several users are now becoming frustrated by the constraints of being tied to the central computing facility and are looking at alternative options that give the divisions and operating companies more control over their local reporting needs.

## Distributed consolidation

A more recent computerisation activity was undertaken by a major conglomerate operating in the service industries. The clear motivation for the project was to devolve the responsibility for the data entry needed to support a company-wide executive information system. It was accepted at the outset that the consolidation was for management purposes, so the minimal amount of inter-company trading was insignificant.

The approach taken was to develop a 'standard' spreadsheet that enabled the profit and loss and balance sheet to be entered and validated locally and reported along with the cash flow statement, which was derived automatically. This standard spreadsheet was tailored to handle product analyses and key performance indicators, which were specific to each operating company. Each spreadsheet generated a report file, which was transferred automatically to head office using a standard communications package with autodial and answer modems. The resulting system relied on the local management teams to initiate their data transfer on the special data of the monthly accounting timetable.

Introduction of this approach benefited both the head office and the subsidiaries. The head office delegated the task of data entry and was assured of receiving a complete report on schedule. It also knew that each company had a basic set of management reports. From the subsidiaries' viewpoint, the system gave a better range of reports than they had had previously and enabled them to use the corporate executive information system to monitor their own performance. They could also be sure that the figures at head office were identical to the ones they had reported.

## Financial modelling

A typical corporate financial model was developed by a large consumer goods conglomerate in the early 1980s. It illustrates several characteristics common to such models. The purpose of the model was to evaluate the implications at group level of likely scenarios derived from a major strategic planning view. In particular it had to assess the international tax liabilities, future profitability, funding requirements and future dividend stream.

The major components of the model were input modules, output reports and the underlying logic. Input modules allowed revised assumptions to be made in several areas:

- divisional trading performance
- capital expenditure programmes
- economic factors, such as interest and exchange rates
- sources of funding and dividend payments.

The logic was very complex, incorporating not just the consolidation to get group figures but also the UK and US tax calculations, assessment of funding requirements based on trading results and investment, the iterative calculation of borrowings and interest payments, and the impact of different dividend policies. The output was primarily tabular but could also be linked into a graphical system for distribution in reports and access using the corporate executive information system. In practice the general assumptions for each run were specified by the directors, but the precise specification and the interpretation of results were carried out by financial analysts.

The results of using the model were that the group executive committee could evaluate various options for organic expansion and acquisition and assess their impact on the portfolio of businesses within the group and their capabilities to meet the expectations of shareholders for growth in earnings and dividends.

## Decision Support Systems

### *History of decision support systems*

The term 'decision support system' covers two distinct application areas, both of which enhance the information available to planners and analysts. The first area concentrates on data retrieval, enabling the user to focus attention on a very precisely defined area of activity. The benefit for the users is that they can assess the impact of a particular policy by separating out those activities directly affected by it. For instance, a brewer selling an alcohol-free lager might be especially interested in targeting promotions on pubs that have car parks. In order to monitor effectiveness of the promotion the brewer would have to identify such outlets and compare the product's sales there with those from the remaining outlets.

The second area of decision support systems includes the entire range of modelling activities carried out by middle managers, typified by answering questions of the 'what if' variety. They are normally limited to a specific corporate function, such as finance, marketing or production, and carried out by the middle managers and analysts within the relevant department. Within this categorisation we can identify two distinct types of model. One is the casual 'back of the envelope' calculation typically carried out by a middle manager during the process of budget preparation: 'what if sales grow by 100% next year?', 'what if we achieve 25% market share?' and so on. The other type of model is constructed by a specialist and provides a more complex description of the effects being investigated. In the marketing area, for instance, much research has gone into assessing the impact of advertising, and many companies have sophisticated models using complex statistical techniques to estimate such effects and produce forecasts incorporating them.

The previous section has already paid some attention to financial models, because they tend to use the same software tools as consolidation systems. In most other areas, though, the tools are different for one of two reasons:

- The general 'simple' models are normally constructed by non-specialists and may well be used once only. Hence, the emphasis is placed on ease of use and speed of rule definition. Complexity of calculations is normally sacrificed to enable such use. The typical example would be the ubiquitous computerised spreadsheet, which must be the most widely used decision support system.
- The specialist models are similar to financial models in that they require more complex rules and are most stable, but the rules they require are different from those normally found in consolidation systems. For instance, most approaches to market modelling entail sophisticated statistical methods, and at least one calls on mathematical techniques originally developed to describe nuclear physics!

Specialist decision support systems came of age during the 1970s in large time-sharing computer bureaux, largely because the work to be carried out did not meet the typical data processing requirements of predictable, well-defined tasks. Decision support is by definition somewhat open-ended and normally entails urgent deadlines.

In the early 1980s the growing availability of microcomputers gave the analysts the self-contained tools they needed, and the advent of spreadsheets enabled the more straightforward 'what if' calculations to be transferred from the calculator and paper on to computers. For the first time, users were provided with an environment where the marginal cost of further analysis was zero. However, there is now a growing trend towards closer links to mainframe computers, fuelled by the easier availability of data from 'data warehouses' held in the large company mainframe. However, most of these approaches continue to use the local power of a PC to carry out the analysis, supplemented with the capability to down-load selected arrays of data from the mainframe database.

## *Types of decision support system*

As the previous section has implied, the range of different types of decision support system is very wide, and this chapter can do no more than give examples of some common techniques that have been applied. This section outlines a few such examples.

### Spreadsheet

This approach is probably well known to readers. The PC-based spreadsheet is surely the most widely used decision support system because it is so simple for computer non-specialists to use, designed as it is to mimic the accountants' paper spreadsheets.

The single most important characteristic of the spreadsheet is that it is generally written and maintained by the end-user. It also has the virtues of instantaneous recalculation of rules whenever data are unchanged and of being easy to define. It does hit limitations in terms of data volumes that can be handled, complexity of rules—especially conditional calculations—and the problems of providing clear documentation and ease of support for other users. The very simplicity of the specification avoids the traditional systems analysis approach, which, despite its other shortcomings, does at least encourage production of documentation *en route*.

Spreadsheets are used for a wide variety of activities, with two main areas of application—storage and retrieval of information, and carrying out calculations. In both cases they tend to act as a personal system for the user concerned, rather than as a corporate facility. As a storage medium, the size constraints and the awkwardness of moving from one spreadsheet to another restrict use (although these are being overcome). As a tool for calculations they offer a flexible, straightforward means of systematising many of those tasks, that are frequently repeated, either on a formal basis or as 'back-of-the-envelope' calculations for personal use. In both areas, the spreadsheet gives the user control of the system so that it can remain confidential and is not affected by other computing priorities.

## Multi-dimensional and relational databases

Probably the next most common decision support system is based on the idea of an underlying multi-dimensional database of information. A typical marketing application might analyse sales by period, region, channel of distribtuion, product line, pack size and perhaps some other criteria. Market planners can extract any defined portion of this structure and so analyse what the impact of current policies might be; for instance, they might compare sales of a chosen product across several channels of distribution to see the effect of some promotional activity.

A relational database provides the capability to link tables that describe different reporting items. These tables may correspond to the dimensions described above. For instance, a product table may hold the product price as well as show which production lines are used for its manufacture. A relational database would tend to hold a lower level of aggregation than a multi-dimensional database, but the nature of the enquiries made from either system would be similar. Most multi-dimensional systems contain some relational capability, so the differences are largely of scale.

In both cases, the main benefit is that management users can make more detailed requests for information from the corporate databases without need to refer back to the data processing department for special programming. Use is concentrated outside the data processing environment, so the system must be easy to learn to use. This has led to the development of access using an 'English-like' command language, which planners and analysts find more powerful than a menu-driven approach. This does restrict use to the middle management level and to planners and analysts. The output tends to be numeric, although most software systems can offer limited graphics, and some have more complete graphic capabilities.

## Specialist modelling systems

The generic approaches described above may be supplemented by a variety of specialist systems normally operated by functional specialists within the organisation. Increasingly these are being branded as 'expert systems', as they encapsulate specialist knowledge about the function concerned.

For instance, the typical sales and marketing department of a fast-moving consumer goods company could operate a variety of such systems:

- The sales forecasting system would be operated on a routine basis and would probably be linked into stock control and production planning activities. The underlying statistical methods would probably include trend analysis and seasonality estimation, but the system would also have to incorporate management inputs on the likely impact of planned promotional activity.
- A market model would probably be used to monitor the effectiveness of advertising, price and other promotional activity on brand share, for both the company and its competitors. This would probably be used on a diagnostic basis to understand historic market movements, and in a predictive mode to evaluate the likely impact of planned tactics.
- A special-purpose new product model might be available to view the uptake of a newly launched brand, as potential consumers move through various stages of brand acceptance and awareness, intent to purchase, first trial and regular use.
- Decision trees and similar decision analysis tools could be used to evaluate the results of market research, product test and test marketing; typically they would enable the department at any stage to drop the project, continue testing or commit to full product launch.
- These systems would probably be complemented with a large multi-dimensional database, analysing sales by month, product, region, channel of distribution and so on, of the sort described earlier.

Different functional areas will obviously have different needs and specialist tools. One area that has been particularly effective is the use of visual simulation systems to aid production planning. The special characteristic here has been the display of the model's output as a picture summarising the production process. The production scheduler is able to 'run the clock forward' to see the impact of a proposed schedule. If the results are unsatisfactory then a new schedule can be evaluated.

Specialist modelling systems are less generalised than other decision support systems, so the input tends to take the form of an input screen of the parameters required. Increasingly they are fed electronically from underlying computerised sources of data. Because the range of options needed is fairly small, such systems can be made easy to operate, but their outputs tend to need interpretation from an experienced user of the particular model in question, normal at planner/analyst level.

### Benefits of decision support systems

While there are clear differences between the types of decision support systems described above, they tend to offer the same types of benefit to the user. These benefits all arise from using the information available to improve the managers' understanding of what is happening, both within the organisation and in the external environment.

The first benefit is an improvement in the quality of the tactical decisions made by the organisation. The 'typical' marketing department described above would be looking to improve its performance in the areas of brand management and new product development. By assessing the relative impact of different activities, the marketing team could assess not only how much money to spend on promotional support, but also how to allocate the budget most cost-effectively. In one of the most famous case studies, an American brewer found that it could reallocate its marketing budget and save enough money to build a new brewery every year, without damaging its sales!

The second benefit is a better appreciation of the underlying relationships that influence an organisation's performance. Economists have long preached the theory of price–demand equations, and similar relationships can be found affecting all areas of activity where the number of consumers is fairly large. It is often claimed that the process of undertaking a modelling exercise provides as much benefit as the results themselves. This reflects the observation that those closely involved begin to learn how the organisation responds to the different policies that could be adopted.

The third area of benefit is the capability to evaluate the implications of a given set of assumptions. This provides the ability to develop strategies that are robust to different external factors. For instance, a chemical company might find an optimal policy based on, say, its best estimates of likely future crude oil prices. However, if this policy is particularly sensitive to this assumption, it may be better advised to adopt an approach that provides a lower expected return, but that is less dependent on such a highly volatile price. In summary, this use of models should enable management to use the power of the computer to test out their plans much more thoroughly than was conceivable without such systems.

### Decision support case studies

#### Multi-dimensional database

A typical multi-dimensional database was developed for a major brewer in the mid-1970s. The system was used to monitor data on sales volumes from two sources—the internal company sales statistics and the reports from an industry association which collated returns from all the major companies.

The first (internal) database was analysed by period, region, channel of distribution, product and mode, i.e. whether the figures were actual or budgets. Based on internal figures there was a detailed product analysis, including individual brands and pack sizes. This system was used primarily for management reporting within the marketing department. The second database was based on industry-wide figures so that the range of data coverage was determined by what was agreed by the industry association. The dimensions were period, area, distribution channel and product group. The main use for this system was to monitor national trends to assess the extent to which the company was out-performing the market, not only in terms of overall sales volumes, but also in developing new product types and packaging options.

The development of the system showed several characteristics that are typical of such systems. Initial work was carried out on a package that gave powerful capabilities for rapid development, responding to feedback from initial use of the system. However, this option became very expensive once this initial development had been carried out, so a bespoke system was written focusing on the facilities of the original package that were actually used. This transition did not realise savings on the operating costs, but did enable about four times as much work to be carried out for the same budget. Both these systems were operated on external time-sharing bureaux, and eventually the company decided to transfer the system to a departmental minicomputer, allowing the marketing department to expand its use still further at zero marginal cost. This formed the basis for an expansion of data coverage so that analyses and reports could be requested down to individual outlet level.

As with most such systems, it is difficult to identify specific savings arising directly from their use. However, this database provided the first means by which the company could identify the rapid growth of the take-home sector, and this realisation led to the establishment of a special operation to handle this trade.

## A diagnostic marketing model

A more specialist approach was developed to interpret the mass of data provided to a food manufacturing company from the retail audit process. The underlying bi-monthly data showed levels of sales, price, stock and distribution analysed by product, TV region and type of outlet. Reports were also available showing the advertising spend by product by region.

The model enabled the marketing department to work out where the major problems actually lay, bearing in mind that a large loss of market share in a small region might be less significant than a smaller loss in London and the South-East. It also enabled them to work out how much was attributable to price changes, how much to changes in distribution and how much to consumer pressure. These could in turn be related to promotional and advertising pressure. After some months' experience the marketing team was able to calibrate the model to estimate the impact of advertising and price changes. They could then use the model to evaluate different marketing strategies and to monitor new advertising campaigns to ensure that there was no drop-off in response.

The benefits were achieved both in improving the effectiveness of the marketing spend and in increasing their awareness of the dynamics of the market-place for their product. As a result of this learning they were able to isolate the different effects of brand advertising on the brand itself, the product sector and the market as a whole. After developing experience with the model, they re-launched an existing brand with substantial marketing support and it grew from 5 to 15% market share in a growing market, while retaining its price and profit margins.

## Executive Information Systems

### *History of executive information systems*

Executive information systems are the most recent to emerge of the components of the comprehensive management information systems and, indeed, the precise definition of an executive information system, and its distinction from a decision support system, remain somewhat unclear. Certain characteristics are universally agreed:

- An executive information system is used personally by the most senior managers of the organisation, normally the executive directors in a company. It is used primarily as a general management tool, rather than to support one specific function.
- It is specially designed so that the hardware and software combine to make the system very easy to use by people who have neither computer experience nor the time to learn.
- It uses well-designed graphic displays to enable the information to be readily interpreted.
- It provides wide-ranging coverage of the key information needed to control the organisation and to develop future strategies.
- More recently the following additional attribute is being recognised as an important element of a true executive information system. Its users are predominantly concerned with the general issues of running the business and use it to communicate across functional and departmental barriers, rather than simply as a specialist tool.

This definition excludes specialist systems that are developed to meet a particular functional need, however important that may be to the organisation. For instance, several companies have developed a user-friendly interface, allowing somewhat limited access to, say, their marketing database, so that it can be used by the marketing director. A true executive information system would normally be expected to encompass data from financial and production departments alongside the marketing data, and indeed it is this use to bridge such departmental barriers that distinguishes them from other systems.

The earliest executive information systems were developed in the late 1970s for several large UK and US corporations. Most of these early systems were developed explicitly to meet the needs of a single product champion. They all encountered the same fate—namely a lack of interest from other potential users—and consequently failed to survive once their sponsor had left to a different post. One of the first systems developed with the explicit aim of serving an entire management team was developed at Imperial Group between 1981 and 1986.

Following this experience the first commercially available executive information systems became available in the UK and US in 1984 and since 1986 there has been a growing interest in the area with several companies announcing new products. More importantly, there is also a growing body of experience with such systems along with case studies of the benefits that can be achieved and lessons for the implementation process involved.

## *Categories of executive information system*

Over the past few years we have seen the development of three distinct approaches to executive information software. The first category is those developed in-house to a customised specification. Most of these are additions to an existing system to provide a limited easy-access viewing capability. Some systems have been developed to meet unusual characteristics of the business itself. British Airways, for instance, has built a system designed to match the network nature of its operations. An in-house development places considerable pressures on the development team, not just to complete the software development in the short time scales demanded by senior executives, but also to produce an effective specification based on the inevitably limited access made available by such managers. Consequently, most effective executive information systems have been based on the growing base of commercially available software in this area.

Most of the software systems provide a development toolkit, which enables an in-house programming team to develop an executive information system; the most widely used systems in this category are FCS-Pilot[4] and Comshare's Commander[5]. This approach permits a user organisation to incorporate within its system a wider range of options than may be provided by an executive information package. Whether this is a benefit or not depends on several factors, including the development times and costs and, more importantly, the ability of the in-house team to assess directors' requirements based on the small time available for such discussion.

The third category of system is a packaged executive information system. In this case the system is ready for use, and requires only that the client organisation defines its structure and transfers the relevant data from the underlying systems—often the consolidation system described earlier. The packaged approach enables a small internal team, often based in the finance department, to create a working system in a matter of a few weeks, so that the directors can rapidly see the results once they have approved the initiative.

Other advantages are that the package already incorporates many facilities found useful at director level within other organisations and that the system is designed to handle structural changes without requiring any programming effort. However, the main advantage is that the packaged approach enables a directors' information project to be controlled by the information users and so to reflect their constraints and priorities. At present there is one widely used packaged executive information system, namely Resolve[6], supplied by the executive information consultants, Metapraxis.

A somewhat different type of system is closely related to these executive information systems. This recognises that a large number of senior-level decisions are made by groups, rather than individually. This identifies the requirement for a boardroom display system, which allows a wide range of media to be controlled by an executive during meetings. The main software dedicated to this area is Vision[7], again provided by Metapraxis. This enables such sources as television, videotape, 35mm slides, video-conferencing and external databases to be controlled by a director using a 15-key infra-red controller, alongside whatever executive information system the organisation uses.

## Uses of executive information systems

Over the past four years a growing body of experience has developed about how executive information systems are best used, and what distinguishes them from more general decision support systems. The first factor has been mentioned above, namely that a major aspect is their use by groups of executives. This use may be in formal board and committee meetings, in informal review meetings in an executive's office, by two executives communicating over the telephone, or even via messages and memos.

Frequently the benefits arise not so much from the information itself—this should be well understood by the analysts using more conventional tools—but from enhancement of directors' capability to communicate and convince their subordinates of the importance of certain issues. An alternative name is corporate control system, and this description encapsulates more precisely the nature of the system. It provides the backbone to the review meetings held between the director and subordinates, which provide the main means by which a director can effect action within the organisation.

This is reflected in four distinct styles of use of executive information systems:

- Analysis. This is most often undertaken by the finance director or his or her team and is the process of tracking down the major problems facing the organisation, diagnosing what is going wrong and identifying who is responsible for corrective action. Three techniques have become particularly important—trend analysis, 'drill-down' or variance analysis, and exception reporting.

- Browsing. This is similar to analysis but less well structured. It tends to take the form of a director undertaking a personal briefing session to find out what is happening in a particular part of the organisation.
- Convincing. Having found a problem or an important issue, the director has to convince the manager responsible that the latter should take appropriate action. This activity may be a presentation at a board or committee meeting, or may take place less formally.
- Data reference. Here the director merely wants to know the relevant figures. The most commonly cited example is the company's share price.

There are two distinctive features of the directors' role in an organisation. First, their responsibilities force them to be generalists rather than focusing solely on one particular function. Second, they are responsible for formulating policy rather than for carrying it out themselves. This means that they use an information system for different purposes from the middle managers. In particular they tend to use the system for persuasion—the 'convincing' role outlined above—and to assess the performance of their subordinates. They are less involved in modelling, leaving that to the analysts who have the time, the specialist expertise and the detailed knowledge of the relevant assumptions that have to be made. They will, however, frequently view the results of models that have been run elsewhere.

On the practical level, executive information system use is distinguished by an ergonomically designed interface that allows directors to use the system with little training. Many systems have used touch-screens as the control medium, but with more experience it appears that these are more impressive in demonstration than in practical use: prolonged use leads to arm-strain, and they are impractical for use in boardrooms and committee meetings. Instead, users are increasingly turning to the use of an infra-red keypad, which is as familiar to use as their telephone or calculator.

## *Benefits of executive information systems*

As indicated above, the primary benefits of an executive information system derive from improvements in the overall management reporting process. Often, indeed, the implementation of such a system forms an essential part of a shift in management style, often paving the way for increased delegation to the divisions and operating companies. However, there are also several specific areas where the early users of executive information systems have identified major benefits. These have often been readily quantifiable and have paid for the investment in implementation many times over.

- Better budgeting. Well-implemented executive information systems have almost always highlighted areas where the budgeting has been slack, in terms of either the overall targets or the monthly phasing.
- Earlier detection of problems. The executive information system normally

provides the easiest way to track and diagnose problems, often identified initially by the system's exceptional reporting capabilities. Often a vital period of two to three months can be gained, enabling a problem to be addressed before it gets completely out of control. Because the issues covered are always important, the profit savings can amount to several million pounds for any one problem.

- Quicker agreement on key issues. The clarity of well-designed charts enables the discussion at key meetings to focus on how to tackle problems, rather than on debating whether a problem actually exists. Availability of the system at board and committee meetings ensures that the main topics can be discussed immediately instead of waiting for the data to be provided at the following meeting. In both cases, this results in better strategic decisions, based on a clearer understanding of the business issues.
- Assessment of managers: the directors have few resources at their immediate disposal and the most valuable one is the management team itself. The executive information system enables the director to assess how well each manager is running his or her operation and to ensure that the best people are in the most important jobs.

## *Effective implementation of executive information systems*

More than any other of the systems outlined above, an executive information system depends for its success on the way it has been implemented. The reason is simple: the main benefits derive not just from the information made available, but from the personal use of the system by directors, and there is nobody in the organisation who can instruct the directors to use it. This situation is aggravated by the fact that few directors have any experience of using computers and most have little time in which to learn.

Over the course of about 80 executive information projects with major organisations, Metapraxis has identified some factors that are essential for a successful initiative:

- The project must have an active sponsor at director level, who should personally review project progress on a regular basis. Without this firm orientation, implementation exercises lose the sense of urgency that is needed and tend to get bogged down in technical issues.
- Review meetings should be held at least every month, and each meeting should be centred around live use of the system with real up-to-date data. This implies that the implementation team has about a month to create the initial prototype with real data; this factor alone argues strongly against an in-house customised approach.
- All data shown to users must be accurate and up to date. Directors cannot evaluate the system on the basis of demonstrations that they are shown; they will only understand its impact on their management style when they begin to use the information as part of their job. This means that they must

be able to act on the results, for instance by faxing a paper copy of a chart to a subordinate and discussing it on the phone. As a rule of thumb, we have found that over 80% of the time spent on implementation should be spent addressing the data requirements. The technical issues must not be allowed to dominate the project.

- The system itself and the implementation process should be controlled by the information users, frequently within the finance team. Only they have a full appreciation of the changing business needs of the organisation. You should expect the structure of the system to evolve to meet the changing needs of every large organisation. This evolution must be so straightforward to handle that the executive information system is the first area where changes are reflected.

## *EIS case studies*

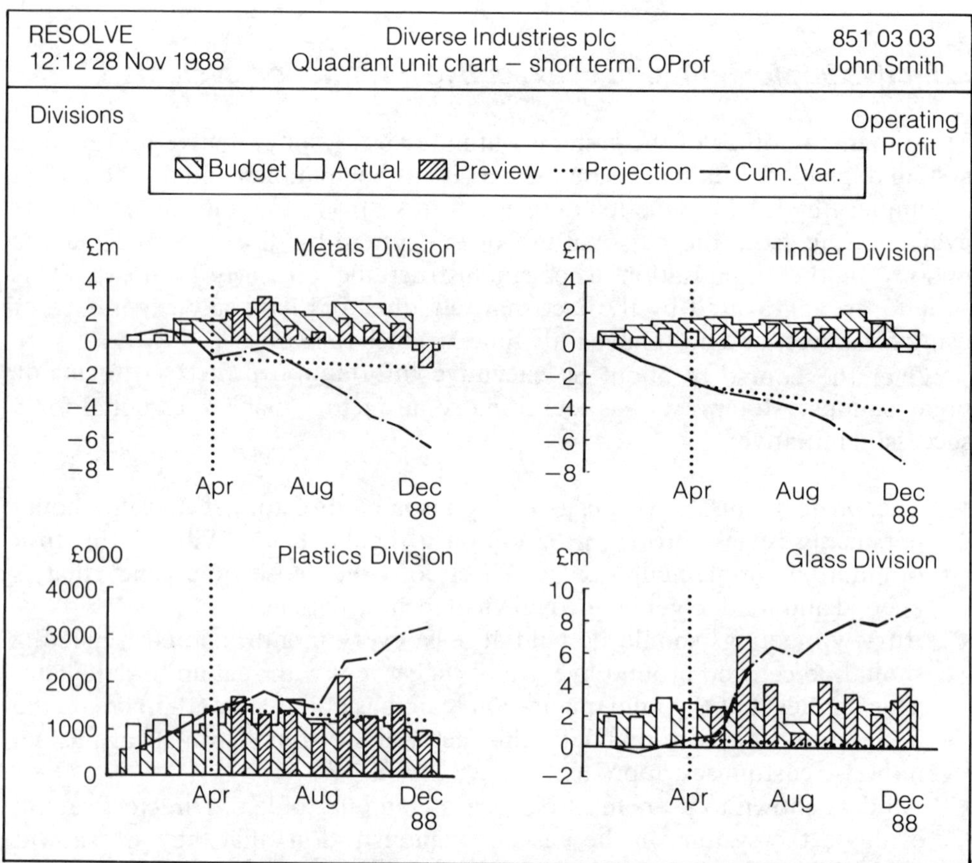


**Figure 2** Trading performance of the four divisions of Diverse Industries.

## Examples of output

Because an executive information system is by definition highly visual, the best way to convey how it is used is by giving some examples. Figures 2–5 are based on data from one of the early UK users.

Figure 2 shows the trading performance of the four divisions of Diverse Industries. We can see that the major problem lies in the Timber Division, but that these problems are long-standing. However, the Metals Division has hit a sudden crisis that needs immediate attention. Figure 3 shows that the problems in the Metals Division stem almost entirely from the Northern Castings operation. Figure 4 looks at the margins within Northern Castings and shows a major write-off at the previous year-end followed by a similar result in the current month.

In the original company, this was identified as a major problem with the stock control procedures in the operating company. The executive information system could not cure those problems, but did enable them to be rapidly identified, leading to an agreed course of remedial action.

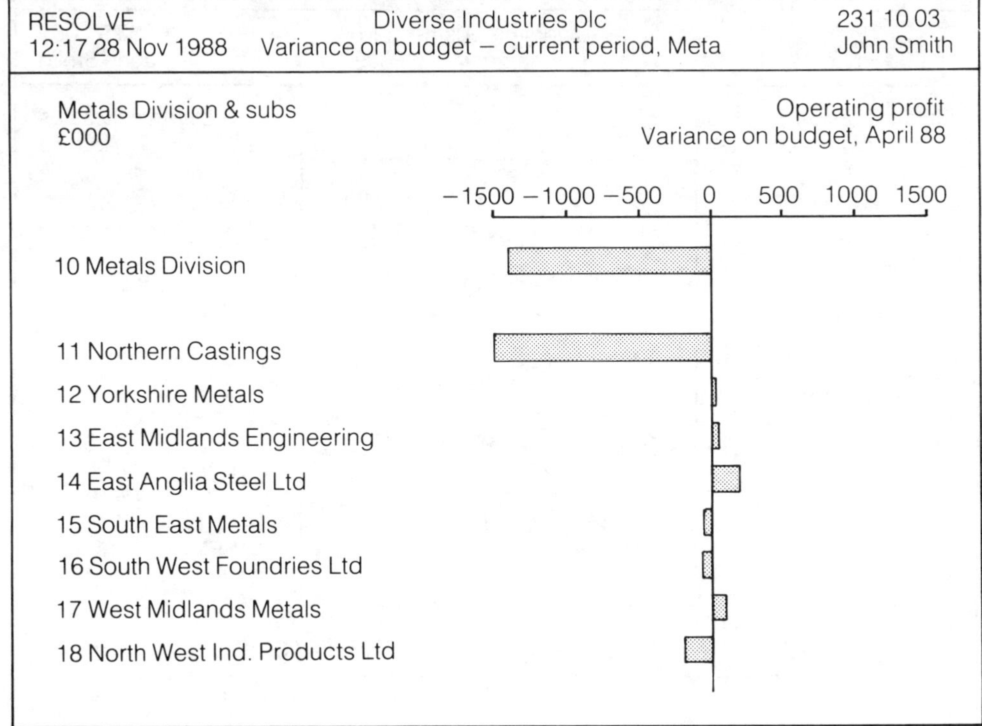


**Figure 3** Operating profit or loss for various sectors of the Metals Division.

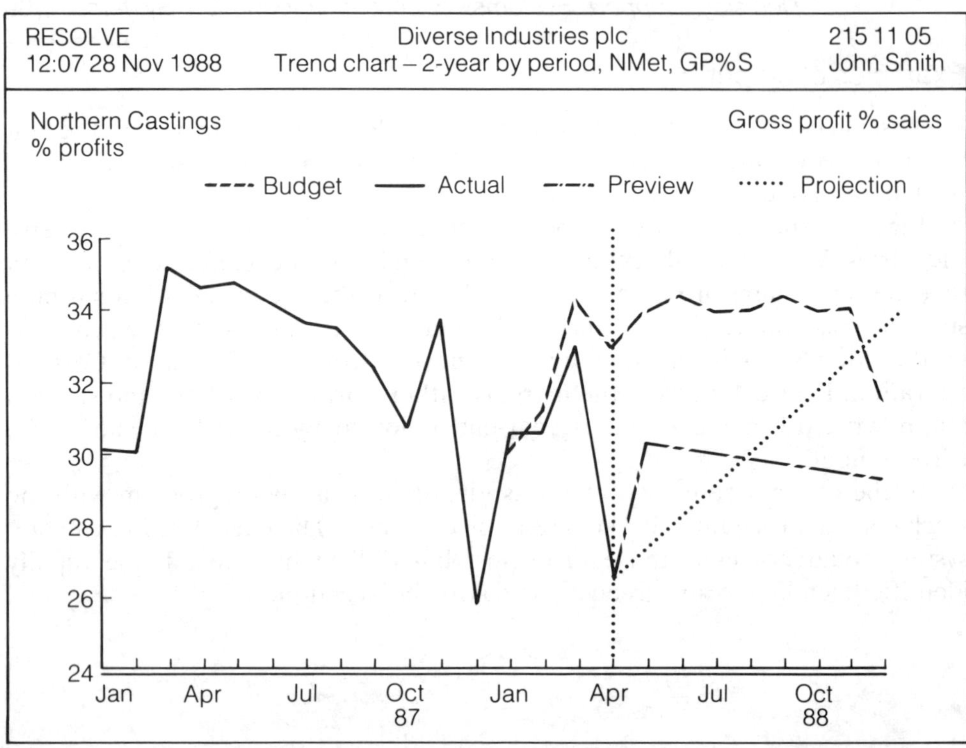


**Figure 4** Margins for Northern Castings.

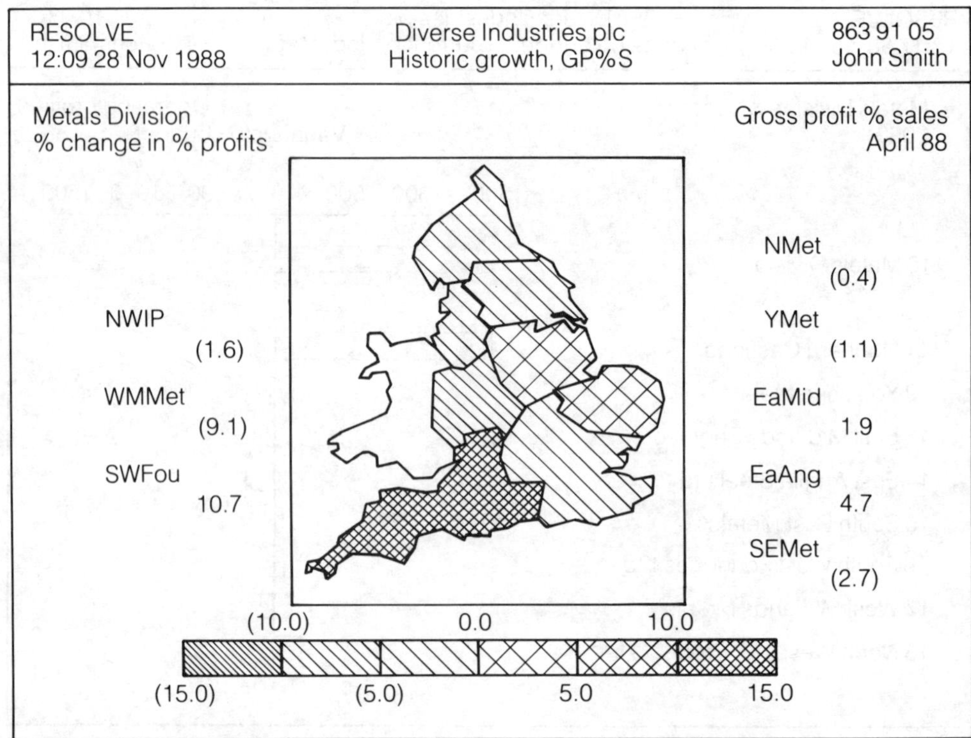


**Figure 5** Historic growth of the various sectors of the Metals Division.

### The budgeting problem

Another effective use of an executive information system was demonstrated by a large American consumer goods manufacturer in the early stages of its use of the system. Sales in a chosen division had been declining, but the advertising budget was held, not because the marketing team had a positive strategy to rebuild sales, but through inertia as no decision was taken to reduce it. This issue was identified by use of the executive information system, leading to a swift agreement to reduce the advertising. A saving of several million dollars was made within the first two months of starting to implement the system!

## Effective Implementation of Management Information Systems

In this chapter I have examined a wide range of different types of management information system, and I have attempted to distinguish the different roles that each has to play. Ideally every large organisation would have a fully co-ordinated combination of such systems, enabling local entry and validation of data at the relevant operating unit, with automatic electronic links into a 'data warehouse' from which the numbers can be extracted by analysts for modelling exercises using their decision support tools, and for distribution to directors via an executive information system.

Of course the real world is not so straighforward, and many organisations are faced with a more fragmentary situation, where few of the base systems are in place and the computing resources at different locations may use different software and incompatible hardware. Faced with these difficulties, many organisations are attracted to the option of a 'bottom-up' systems review, working on the logic that it is not feasible to implement effective decision support and executive information tools until the lower-level feeder systems are in place.

Normally, this type of systems review includes the following steps:

- User requirements. Users at all levels are asked to define their needs and a series of interviews is undertaken to elaborate on these.
- Systems specification. The main components of the proposed system are identified and the various database contents and interfaces are defined.
- Software evaluation. Different software packages, including the alternative of a bespoke development, are evaluated and rated according to the technical specifications derived earlier.
- Systems development. A major exercise is undertaken, leading to a gradual implementation of systems from the bottom upwards, each system feeding off the data already computerised. At each stage a parallel run is carried out for two or three months to ensure compatibility with the existing techniques.
- Review. When the entire system has been completed, a review is carried out to assess any areas of shortcoming.

In theory, such an approach will result in a well designed and developed system that successfully addresses the stated needs of all levels of user, as defined at the outset of the project. In practice, however, things are rarely so simple and this conventional approach builds in two major contraints that are liable to prevent a successful project.

First, the users' requirements may be impossible to define. The routine aspects, such as consolidation, may be straightforward and amenable to this approach but the directors' needs from a system are considerably harder to address—they rarely know what their 'needs' are, and so the specification for this area is at best imprecise. Second, the time scales involved in delivering a workable system become protracted—normally to at least 12–18 months. Not only does this wait delay the anticipated benefits from the project, but it also means that the users' requirements from the system will have changed. The combined effects are that the most senior users, whose decisions have the greatest impact, are exactly those whose requirements are least well specified at the outset and who have to wait the longest time for a working system. By this time it is highly likely that many of the personalities or their expectations of the system will have changed. Of course it is then too late to alter the underlying systems, so the final composite system is unlikely to achieve the benefits that would arise from effective use at senior level.

An alternative approach to an integrated implementation has been growing in favour among consultants with more experience in implementing systems for use by directors. In this approach, the emphasis of the project is shifted towards the needs of the information users, particularly those at the highest level. However, it is explicitly recognised that they have neither the time nor the technical expertise to define what they need. A preliminary executive information system is built in a few weeks, concentrating on a chosen part of the overall organisation. This system is used for a couple of reporting cycles, so that the directors can gain direct exerience of how their own work is affected by the information now available. By limiting data coverage the preliminary system can be created without needing a complete computerisation of all underlying systems. As the executives' requirements evolve in the light of initial experience, it becomes clear what demands are likely to be put on their analysts for modelling, so the specification of the decision support systems begins to crystallise. This in turn affects the requirements placed on the consolidation system.

By placing greater emphasis on how the information is going to be used by decision-makers, the project remains firmly linked to the business needs of the organisation, rather than acting to constrain the types of data and when they become available. The net result of this change of emphasis is that the project is no longer a major act of faith, in which the project's sponsors have to hope that the final system will be deliverable at the end, but have no usable system in the meantime. Instead, there is an evolution of systems, gaining top-level commitment and approval at regular intervals, and based at each stage on feedback from use of the system in real-life business situations.

## Summary

This chapter has described the main components of integrated management information systems and assessed their impact on the operation of a large organisation. While it is clear that the various specialist and functional tools may well vary from one type of organisation to another, as the use extends to more senior managers so the emphasis changes from looking at events to judging people and their effectiveness.

This means that the natures of such systems at director level become quite similar, and that there is considerable benefit to be derived from basing one's own system on the collective experience of other trail-blazing users. The best executive information systems have incorporated such experience through devices such as ergonomically designed charts and exception reports, so much so that an identical system has been found effective across many different types of organisation, including industrial and commercial corporations (in areas such as manufacturing, retailing, distribution and service industries), financial institutions (banks, insurance companies, etc.), the public sector (including government agencies, nationalised industries and health authorities), national economies (as an aid to ministers and industrialists in Third World and developed countries) and finally for management control of major projects.

It is always easy to be carried away by the undoubted technical challenges of addressing the management needs of a large organisation and to lose track of the main reason why the exercise is being carried out—namely to improve the effectiveness of the organisation through better management and decision-making. I am sure that we are all familiar with the frequent apology for missing or inadequate information: 'Well that's all we can provide because it's been put on the computer!'

The only way that we can ensure that the next generation of information systems proves more effective than their predecessors has to be by placing the overall control and prioritisation of such systems firmly in the hands of the users of the information. The tools are there already but few organisations have developed an infrastructure that allows the managers to use them effectively.

## Notes

The notes provide addresses from where further details of the products can be obtained.

1. FCS, Thorn EMI Computer Software, 4–7 Station Road, Sunbury-on-Thames, Middlesex TW16 6SB, UK.
2. System W, Comshare, 32–34 Great Peter Street, London SW1P 2DB, UK.
3. Micro Control, CAP, Barrington House, Heyes Lane, Alderley Edge, Cheshire SK9 7JZ, UK.
4. FCS-Pilot, Thorn EMI Computer Software (address above).
5. Commander, Comshare (address above).
6. Resolve, Metapraxis, Hanover House, Coombe Road, Kingston-on-Thames KT2 7AH, UK.
7. Vision, Metapraxis (address above).

# 3.3 *Artificial Intelligence and Expert Systems*

*Soundar R. T. Kumara, Rangasami L. Kashyap and Allen L. Soyster*

This chapter will deal with the concepts and some of the applications of artificial intelligence (AI) and expert systems. It will provide a semi-technical introduction to these two related topics and will indicate the actual and potential relevance of the new areas to managers.

## Artificial Intelligence

*Webster's Dictionary* interprets intelligence as 'the ability to understand new or trying situations'. The more commonly accepted definition of intelligence is 'the ability to perceive, understand and learn about new situations'.

The human brain is equipped with an enormous potential to perceive, understand and learn. If this ability could be duplicated in a computer system, then by the definition of intelligence, the computer should be classified as intelligent. Even if we succeed in duplicating the human reasoning process in the computer, how can we test this claim? The answer can be found in the test developed by Alan Turing. Figure 1 represents this classic experimental situation. The experimenter is connected to the computer and a subject is stationed in a different room. The experimenter interacts with both the human and the computer. If, at some point in the dialogue, he or she cannot

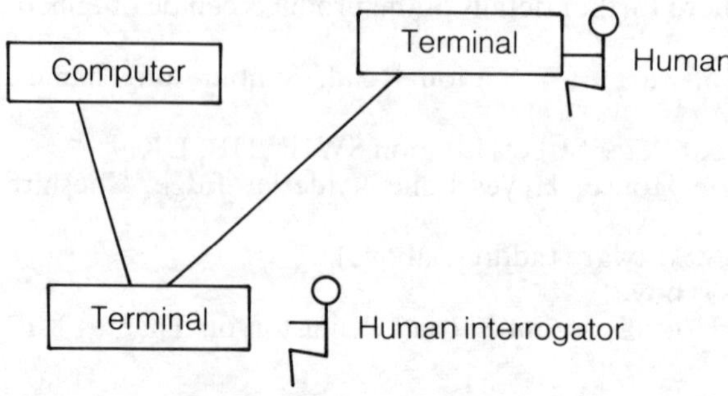


**Figure 1** Turing's test.

differentiate between the computer and the subject, then one has to agree that the computer has performed at the level of the human being. If we say that the human is intelligent, we have to agree that the computer is also intelligent. When human intelligence is captured by an external system, the system is said to based on artificial intelligence (AI).

There are many formal definitions of AI. As this is a relatively new discipline, it means different things to different people. The following are some of the definitions given by different experts:

- Artificial intelligence is that field of computer science concerned with designing intelligent computer systems, that is, a computer system that exhibits the characteristics we associate with intelligence in human behaviour (Barr and Feigenbaum, 1981).
- AI is the science of making machines do things that would require intelligence if done by men (Minsky, 1975)
- The field of AI has its main tenet that there are indeed common processes that underlie thinking and perceiving, and furthermore that these processes can be understood and studied scientifically. In addition it is completely unimportant to the theory of AI who is doing the thinking or perceiving—man or computer (Nilsson, 1971).

The underlying idea in all these definitions is one of understanding human cognitive processes and modelling them on the computer such that the computer can solve the process in the same way as the human. Let us examine some of the basic areas of applications of AI.

## *Applications of artificial intelligence*

We will now identify some of the applications of AI that may be of relevance to managers.

### Natural language processing

One of the application areas that is of immense interest to AI researchers is natural language processing. One of the earliest results, and also a widely used tool, is the augmented transition network (ATN). All natural language understanding systems begin with a dictionary of words associated with the semantics of the sentence. The first phase of most natural language systems is to understand the syntax and validate the construction of the sentences using rules of grammar, usually via an ATN. The second phase deals with the semantics of the sentence. The major uses of natural language processing are:

- answering queries posed in English from databases
- translating a language into English
- comprehending text.

A related application deals with the understanding of spoken language. Speech recognition and speech understanding are among the most important offshoots of AI.

### Robotics and vision

Here there are three major application areas:

- manipulating robotic devices
- planning optimal paths
- sequencing tasks for goal accomplishment.

Currently, systems for welding with the aid of robotic devices and vision sensing are available. In the industrial robotics realm, a current focus of research is on making the robot pick a specific item from a bin of randomly arranged items. The major issue related to mobile robots is concerned to navigating the robot among obstacles. When there are fixed paths, the robot can be guided with the help of wires. However, if the environment is unfamiliar and full of obstacles, there is a need for intelligent path selection and learning from the previous motion

### Manufacturing

Manufacturing is a fairly rich domain for applying AI techniques. However, most of the current systems in the literature deal with generalised problems. Manufacturing problems can be characterised into:

- planning problems
- design problems
- classification problems
- diagnostic problems

Process planning, facilities layout and scheduling fall into the first category. Conceptual design, which essentially consists of interactive synthesis, has attracted much attention in recent years. It is interesting to note that the idea of AI-based conceptual design has been widely used in the fields of architectural and mechanical engineering. Group technology, based on pattern recognition and symbolic matching, is a typical classification problem. Diagnostics is perhaps the most researched area in manufacturing. Diagnostics and control are grouped together in the manufacturing domain.

## *Approaches to AI*

This section deals with two important approaches to building AI systems. Most of the AI systems can be characterised as heuristic programming or logical reasoning.

## Heuristic programming

Most of the ideas of problem-solving formalisation are derived from chess. Most of the computer programs for chess analyse possible alternatives for a given configuration. On the other hand, based on experience and given the opponent's move, a grandmaster looks at all possible configurations and picks the one that gives the best chance of winning the game. This forms the basis for looking at problem-solving in AI-related research as a series of pattern matching steps.

Most problem-solving tasks involve the use of heuristics. Heuristic programming is essentially the basis of AI. A simple example would be the eight puzzle problem. The objective of the exercise is to arrange eight tiles in a particular order. This requires the application of rules for moving the tiles to the left, right, top or bottom. At any given point in time, the state configuration dictates the use of these rules. Most of the experience-based knowledge can be captured in the heuristic programs. Representing heuristics inspired the development of symbol oriented programming languages.

Some of the major considerations in heuristic programming are:

- types of heuristics used
- means of constraint propagation
- evaluator functions on the solutions generated.

## Logical reasoning

Methods for proving assertions of logic are used in the logical reasoning process. These methods are complete and consistent. The programs have a collection of logical facts in their databases, which can be updated as additional information arrives. Given an assertion, the program uses logical theorem-proving procedures to prove the assertion. The major question is: 'what minimal set of facts will be used to prove an assertion?' AI systems in general use logical deductive or inductive reasoning. Deductive reasoning is the most common.

Consider the following axioms:
All birds have feathers (axiom 1):

$A \rightarrow B$ (A, birds; B, have feathers).

If an object has feathers then it can fly (axiom 2):

$B \rightarrow C$ (C, can fly).

By using the transitivity relationship, which is a form of deduction, one can conclude that birds can fly. In a specific instance, if one observes that a parrot is a bird, then by similar deductive inference it can be concluded that a parrot can fly.

Logical inference is typically a multi-step process. It is possible to derive a conclusion in a single step (parrot has feathers) or in two steps (parrot can fly). While building AI systems, it is possible to separate the axioms and facts.

Axioms can be represented as rules, and the principles of logical inference (*modus ponens, modeus tolens* and universal instantiation) can be used to derive conclusions. Most often the goal is reached via a generate-and-test paradigm. The subproblem-solving is triggered by the distance between the current state and the goal state.

## Search—the heart of AI

In generating the path leading to the goal, the system should be able to travel through the solution space, which is often represented by a tree. Each solution is a path in the tree from the root node to one of the leaf nodes. The number of paths increases exponentially with the number of levels and the number of alternative choices at each level to search. One measure of the intelligence of the system is the effort required to go through the solution space and arrive at the goal state. Search is an important aspect of AI.

The underlying concepts of AI are best understood by examining production systems. All AI systems consist of a central entity, the global database, which is manipulated by operators. The application of these operators is controlled by a global control strategy. The control stategy guides the search through the database. There are basically two types of control strategies:

- irrevocable control strategy
- tentative control strategy.

Whenever an operator needs to be applied to the current state, the control strategy guides the search and helps in choosing the most appropriate operator. In the irrevocable strategy an applicable rule is selected and applied irrevocably without provision for consideration later. In the tentative control strategy, at a given state description an operator (rule) is selected, based on an arbitrary scheme or with some good reasoning, and then applied. The search could proceed further, based on need, and provisions are made to come back to this point to apply some other rule and try an alternate path.

Tentative control regimes are of most interest to us. Tentative control can be classified into two categories: backtracking and graph search. In backtracking, a backtracking point is selected when a rule is applied. During the course of evaluation of the path to the solution, should a difficulty be encountered (dead ends or cycling), the process can revert back to the backtracking point and another rule can be tried. In graph search procedures, several operators are applied simultaneously for a given state and alternative paths are generated. Provision is made for keeping track of these efforts. Graph search procedures, which are used extensively in AI systems, can fall into two categories: blind search and informed search.

In blind search, the control procedure does not make use of problem-related information to guide the search. On the other hand, the problem-related knowledge is used to restrict the search in the informed strategy. The blind search procedures are classified into: depth-first search and breadth-first search, based on the method of generating the next (successor) states from the current state. In all these search procedures, we start with a root node, which represents the initial configuration or initial state description. The operators relevant to the problem are applied, checking for their applicability (preconditions), and successor nodes are generated. The terms used in a simple generic tree search procedure are as follows:

1. Root node. The starting state or initial configuration, which represents the beginning of the tree. (A tree has only one root.)
2. Leaf node. A node of the tree that does not have any children (successors).
3. Goal node. The node that represents the configuration satisfying the goal state or solution.
4. Branch. The link that connects any two nodes (i and j) together: node j is generated by applying an operator (rule) k to the node i.
5. Expansion. A node i is said to be expanded if an operator(s) is (are) applied to it and its successors are generated.
6. Termination. The process of stopping the search procedure, normally when the goal state is reached.
7. Level. The root node in a tree is said to be at a depth level 0. Any other node's depth is given by level = level of its parent node +1.

A generic graph search algorithm for blind search is as follows (Barr, 1982):

1. Consider the initial node $S_0$ and place it on a list of expanded nodes, called OPEN or unexpanded nodes. If $S_0$ is a goal node, a solution has been found. Exit.
2. If the list OPEN is empty no solution exists, hence exit.
3. Remove the first node n from OPEN and place it in the list called CLOSED, which represents the list of expanded nodes.
4. Expand the node n after checking for precondition and applying operators. If it has no children (successors) then go to step 2.
5. Place all successors of n in the list OPEN.
   - If it is a depth-first search, place the nodes at the beginning of the OPEN list.
   - If it is a breadth-first search, then place them at the end of the OPEN list.
6. If a successor of node n is a goal node, then a solution has been found. Otherwise go to step 2.

## Depth-first search

We will consider the example of the pawn and 4 × 4 grid problem (Barr, 1981). Figure 2 represents the configuration of the 4 × 4 grid. A pawn is waiting to enter the board from an outside cell on to any cell in the first row. The objective of the game is to make the pawn reach any cell in the fourth row that

☒ Pawn waiting on an outside cell

| 1 | 0 | 0 | 0 |
|---|---|---|---|
| 0 | 0 | 1 | 0 |
| 0 | 1 | 0 | 0 |
| 1 | 0 | 0 | 0 |

**Figure 2** Pawn problem.

contains a zero in it. The 4 × 4 grid can be represented by the positions (1,1), (1,2), . . . (4,3), (4,4). The pawn can move one cell at a time. At any time the pawn has knowledge about only one row of the matrix.

Moves are subject to certain preconditions. In a simple English-like structure, the rules of the game are as follows:

1. The pawn can move on to any of the cells in row 1.
2. If there is a cell containing a 1 and if the pawn is on that cell no further movement is possible.
3. The pawn can be moved to the cell to the left, right, top or bottom from its current cell.

An AI search problem has three components:

- global database
- operators
- control strategy.

1. Global database. A node represents a configuration. The configuration in turn represents a state description. The goal state is a state description of the desired state, namely the cells where the pawn can be located, (4,2) or (4,3) or (4,4).
2. Operators. In this problem the pawn is moved. It must be noted that a cell contains either a one or a zero. The state configuration of the matrix can be represented as follows:

   (1,1) = 1
   (1,2) = 0
   (1,3) = 0
   (1,4) = 0 . . . etc.

The operators are:
(a) move the pawn to a cell below or above
(b) move the pawn to the cell either to the left or right of the current cell.

The rules can be described in an if . . . then structure as follows:

**IF** the pawn is outside the board
**THEN** it can move to any cell in the first row
**ELSE**
**IF** the pawn is on a cell $(x,y)$ containing a 1
**THEN** no further move is possible
**ELSE**
**IF** the cell $(x + 1,y)$ contains a 0
**THEN** the pawn *must* move to cell $(x + 1,y)$
**ELSE**
**IF** any adjacent horizontal cell(s) contains a 0
**THEN** move any of the cells
**ELSE** it must move to the cell immediately below.

We use a depth-first control strategy and blindly expand the nodes for the goal state. The algorithm described earlier is used to generate state descriptions. In a depth-first search, the most recently generated or deepest node is expanded first.

Figure 3 shows the search tree generated by a depth-first search. $S_0$ represents the initial state, the pawn waiting to enter the board. State $S_{17}$ is the

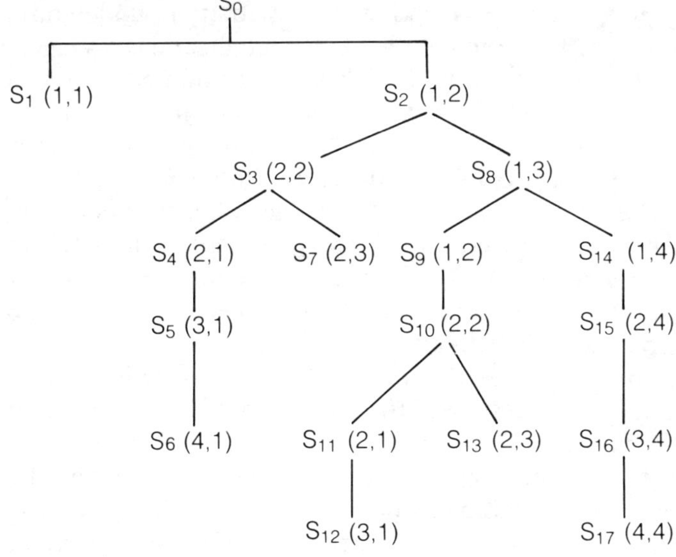


**Figure 3** Depth-first seach tree for the pawn problem.

goal state. The ordering of the numbers corresponds to the order in which the nodes are expanded using the rules to generate children nodes.

### Discussion

The pawn problem would be substantially simplified if the contents of the next row were transparent. We leave the exercise to the reader. If all the rows are known in advance the search is trivial.

The above search methods tend to be exhaustive in nature, leading to combinational explosion. This necessitates the adoption of different strategies that apply heuristic information to the search technique. The search is informed of certain problem characteristics, through an evaluator function that reduces the number of nodes expanded. There are many variations on the same theme resulting in the algorithms A, A*, $A_0$*, etc. The basic notion in all these heuristic search methodologies is one of using the problem-specific knowledge intelligently to reduce the search efforts.

Pattern-matching systems and dictionary-matching systems for natural language or expert systems use a search strategy. The search is either breadth-first or depth-first. In many cases, heuristic search procedures using problem-specific knowledge are employed. Using AI search strategies in expert systems is of value in the industrial engineering and manufacturing domains. The following sections discuss the basic concepts behind the development of expert systems.

## Expert Systems

Expert systems are becoming popular in engineering and management. An expert system can be defined as 'a tool which has the capability to understand problem specific knowledge and use domain knowledge intelligently to suggest alternate paths of action' (Kumara *et al.*, 1988). These are computer programs that combine the collective knowledge of experts in a domain and attempt to function at their level of expertise in problem-solving. The major characteristics of these systems are exhibited by their handling of data and procedures. The declarative knowledge base, consisting of facts about the domain, is separated from the procedural knowledge base. An external inference mechanism (general purpose) activates these procedural rules. Such a separation has distinct advantages because changes in either the declarative knowledge base or the procedural knowledge base do not affect the other. An expert system searches through a large amount of data to generate inferences or reach a solution. If the system needs to be efficient, it should be intelligent in its search through its knowledge base.

Expert systems derive their name from the fact that the systems contain the expertise required to solve specific, domain-related problems. Hayes-Roth *et al.* (1983) categorise knowledge as 'public' or 'private'. Published literature falls into the former category, whereas the knowledge of an individual falls into the latter. In addition, an individual expert can analyse problems in both the exact and the inexact realms, unlike most expert systems. Expert systems try to capture the essence of the expertise (both public and private) related to a problem domain and use it for problem-solving and explanation.

Stefik *et al.* (1982) classify expert systems into the following groups:

- interpretation
- prediction
- diagnosis
- debugging
- design
- planning
- monitoring
- repair
- instruction
- control.

## Components of expert systems

An expert system comprises:

- knowledge consisting of domain-related facts
- knowledge consisting of domain-related rules for drawing inferences
- an interpreter that applies the rules
- an ordering mechanism that orders the application of rules
- a consistency enforcer, when new knowledge is created or old knowledge is deleted from to knowledge base
- a justifier that explains the system's reasoning.

## Constructing an expert system

Building an expert system is time-consuming and non-trivial. The construction of an expert system consists of the following five phases:

- problem definition
- acquisition, representation and co-ordination
- inference mechanism
- implementation
- learning.

With the increasing popularity of AI and expert systems, there are many attempts to build expert systems without considering the basic question: 'does an expert system really help the problem–solving process?. In many cases, it is possible to solve the problems related to manufacturing by direct, and often simple, analysis. In such cases, expert systems may not be appropriate. An expert system may be more effective when:

- problems in the domain cannot be well-defined analytically
- problems can be formulated analytically but the number of alternative solutions is large, as in the case of combinatorial explosive problems
- the domain knowledge is vast, and relevant knowledge needs to be used selectively.

If a problem fits into any of the above categories it may be worthwhile to construct an expert system. The five phases of expert system building are explained in the following sections with the help of relevant examples.

## Problem definition

This phase involves understanding the problem, identifying problem characteristics, outlining the objectives of the problem-solving process and clearly defining the methodology required to solve the problem. This leads to answering such questions as 'what is the knowledge required to solve the problem?' and 'how can this knowledge be acquired?'

## Knowledge acquisition, representation and co-ordination

This aspect primarily deals with knowledge acquisition, knowledge representation, design of inference mechanism, selection of programming tools and knowledge co-ordination.

### *Knowledge acquisition*

Knowledge is defined in *Webster's Dictionary* as 'factors or ideas acquired by study, investigation, observation or experience'. Extending this definition to AI applications, intelligence can be viewed as the ability to perceive the domain through knowledge. In order to have intelligent systems, it becomes necessary to have 'relevant' knowledge. To build the dictionary of knowlege, the analyst constructing the expert system has to acquire knowledge and incorporate it into the system. Knowledge about the problem domain is acquired either from the study of published literature or from human experts in the domain. Facts related to the domain and the specific problem are the components of the 'declarative knowledge base' and rules or procedures that generate paths of reasoning in the expert system are part of the 'procedural knowledge base'.

Typically, knowledge can be classified into four categories (Barr and Fiegenbaum, 1982).

1. Object knowledge. The facts (knowledge) that describe a real-world situation relevant to the problem. For example, 'machines make noise' is a fact about the object 'machine'.

2. Event knowledge. We should also know of events that have taken place or will take place. 'The plant will shut down tomorrow' is the event knowledge about the object 'plant'. It is necessary to develop a formalism for such knowledge to indicate the time course of a sequence of effects and their cause and effect relations.
3. Performance knowledge. This is knowledge about the application of skills or how to do things. It is difficult to draw a line between object knowledge and performance knowledge
4. Meta-knowledge. This is knowledge about the above-mentioned knowledge. It deals with the extent, reliability and relative importance of specific facts, and how they evolved.

Expert systems are designed to preserve the expertise of human experts. One could say that expert systems embody artificial expertise. Human expertise is perishable, difficult to transfer, difficult to document, inconsistent and expensive. In comparison, expert systems are permanent, easier to transfer, easier to document and consistent. The developmental costs may be offset by the system's consistent usage.

## *Knowledge representation*

The knowledge acquired should be represented in a computer implementable form. The representation of knowledge is a combination of data structures and interpretive procedures. Representation schemes are classified by Winograd (1975) into declarative schemes and procedural schemes.

## *Declarative representation schemes*

1. *State space*. This is the earliest formal representation used in AI programs and was developed for problem-solving and game playing. In game playing, each configuration describes a state, 'though strictly state space is not a representation scheme' (Barr and Fiegenbaum, 1982). It can be viewed as a useful means for applying operators, but is not particularly well suited for expert system applications

2. *Logical representation schemes*. First-Order Predicate Calculus is a logical representation scheme. The description of the real world is given in terms of logical clauses. For example, the fact that all humans are mammals can be represented as

"x Human(X)→Mammal(X).

Logical representations are useful in providing formal proof procedures, information retrieval and semantic constraint-checking. They offer clarity and are well-defined and easily understood. Each fact needs to be represented only once irrespective of repeated usage. However, logic schemes offer difficulty in procedural representation

3. *Semantic nets*. Semantic nets were developed by Quillian (1968) as an

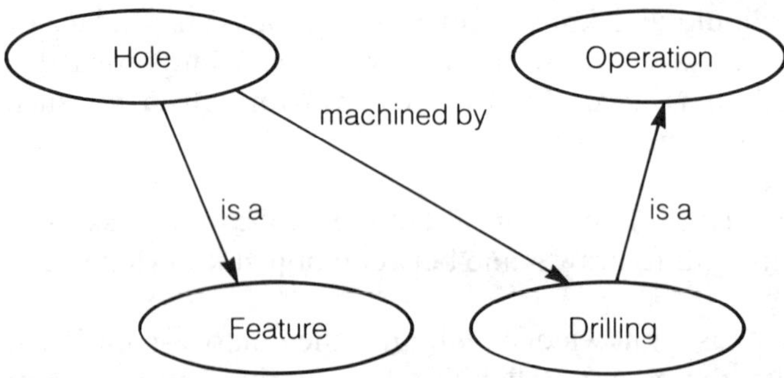


**Figure 4** Semantic net representation.

explicit psychological model for human associative memory. Semantic nets attempt to describe the world in terms of objects (nodes) and binary relations (labelled edges). According to a semantic net representation, the knowledge is a collection of objects and associations represented as a labelled directed graph. A typical semantic net is given in Figure 4, which represents the fact that the hole is a feature that is manufactured by drilling, and drilling is an operation. The nodes represent the facts and directed arcs represent the associated relationships. Semantic nets are easily understood but difficult to implement.

4. *Entity-relationship diagrams*. The declarative knowledge base can be viewed as a relational scheme. The identification of entities, properties and associations (as defined in Date, 1983 and Chen, 1976) is a 'piece of information that describes an entity'. Associations are formed using the entities that define the knowledge base.

5. *Frames*. The notion of frames originally proposed by Minsky (1975) has played a key role in knowledge representation research. A frame is a data structure for representing a stereotyped situation, like being in a certain kind of living room or going to a child's birthday party. Attached to each frame are several kinds of information, such as how to use the frame, what one can expect to happen next and what to do if these expectations are not confirmed. This information is presented in the frame as slots and fillers. Figure 5 shows a frame for a dinner get-together. The name of the frame is 'dinner get together'. The

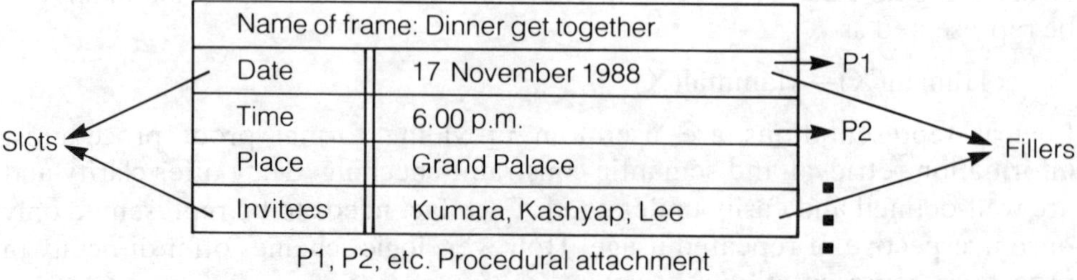


**Figure 5** Frame representation.

slots that represent factual and event knowledge are Date, Time, Place and Invitees. The fillers are the associated descriptions. Frames have procedural attachments. The most commonly used frame representation schemes are FRL (Goldstein and Roberts, 1977), KRL (Bobrow and Winograd, 1977), OWL (Szolovits *et al.*, 1977) and KLONE (Brachman, 1979).

### *Procedural representation schemes*

Declarative schemes deal with facts about objects, events and their relations, and states of the world. Procedural schemes deal with the methods to be used to find relevant facts and to draw inferences. Procedural schemes comprise rules, how to use them and how to build control into the search strategies. All the declarative schemes can be used for procedural representation.

### *Knowledge co-ordination*

Some manufacturing engineering applications are real-time based; that is, the current on-line knowledge from the shop floor needs to be used by the expert system. When building an expert system, it is necessary to consider the knowledge itself of the expert system. It is important to consider the format in which the real-time knowledge is available and to integrate it with the proposed representation scheme of the expert systems. This has led to a hierarchical structure for real-time expert systems. Such systems should incorporate the basic ideas of temporal reasoning. At the lower level we have the real-time shop floor data, and at the higher level a co-ordination engine to integrate on-line knowledge with the knowledge base of the expert systems.

## Inference mechanism

The main purpose of developing an expert system is to generate alternative paths that lead to an inference. In order to accomplish this task the procedures built into the system should act upon the declarative knowledge base in an efficient manner and derive conclusions. The process of generating alternative paths via a reasoning mechanism through the knowledge base to derive conclusions is termed inference. Inference mechanisms are classified by Fox (1985) as

- heuristic search
- analytical tools, like linear programming, dynamic programming and queuing theory
- constraint directed reasoning
- hierarchical reasoning.

Any of these inference mechanisms can be used with the representation schemes discussed in the previous section. The methods of inference can be modelled as rules, semantic nets or frames.

### *Inference via rules*

Rules are typically of the form **IF**<antecedent> **THEN** <consequence>. Some samples follow. A situation leads to an action: **IF** room temperature greater than 197°F, **THEN** change the thermostat setting. A premise leads to a conclusion: **IF** there is no power, **THEN** the computer will be down.

Three aspects of inference mechanism are:

- reasoning through chaining (forwards or backwards)
- rule interpreter
- addition of new rules.

The concept of forward and backward reasoning is explained using a process diagnostics example:

Rule 1: **IF** the square foot weight is less than 12 lb/sq.ft and the tensile strength is greater than 50 lb/sq.ft **THEN** check the line speed of the conveyor.

Rule 2: **IF** the line speed needs to be checked, **THEN** switch the furnace off.

Consider that in a typical situation the following facts are observed.

Fact 1: square foot weight is 10 lb/sq.ft.
Fact 2: tensile strength is 55 lb/sq.ft.

*Forward reasoning* starts with axioms and definitions and makes as many conclusions as possible. The rules in forward reasoning are of the form '**IF** conditions **THEN** conclusion'. Using the rules to reason forward, Fact 1 and Fact 2 can be used in conjunction to infer that the line speed needs to be checked. Forward reasoning, as illustrated by the example, is data- or fact-driven.

*Backward reasoning*, on the other hand, starts with a goal and attempts to find enough facts to substantiate that goal. It is also called goal-driven reasoning. The rules are of the form 'conclusion **IF** conditions', and the inferencing works backward from the conclusion. For backward reasoning the rules are restructured as follows:

Rule 1: Check the line speed of the conveyor, **IF** the square foot weight is less than 12 lb/sq.ft **AND** the tensile strength is greater than 50 lb/sq.ft.
Rule 2: Switch the furnace off, **IF** the line speed needs to be checked.

Backward reasoning can be used to answer the question 'should the furnace be switched off?' In order to answer this question it is necessary from rule 2 to determine whether the line speed needs to be checked. Using fact 1 and fact 2 in the database, the system can conclude that the furnace needs to be switched off. The most important consideration when adding new rules is that of consistency. An expert system should contain a consistency enforcer and a conflict resolver.

The knowledge required to solve a problem consists of three types:

- component knowledge
- constraint knowledge
- procedural knowledge.

The above are explained through a process diagnostic system example. Component knowledge consists of all pertinent information about each process in the overall manufacturing process. This information is represented in the declarative database as facts, and is obtained from the expert knowledge and process parameters. Constraint knowledge deals with the contraints on the selection of each of the processes. This knowledge is represented by rules. Each rule embodies knowledge about the physical constraints associated with a process. This knowledge is derived from the cause and effect relationships of each of the processes. Procedural knowledge consists of knowledge not explicitly covered by the constraint knowledge. It consists of rules to derive the diagnostic conclusions. These rules are obtained by consulting various experts on the shop floor and from process manuals.

## Implementation

The first step in implementation is to decide on the programming language to be used. Most expert systems are based on Lisp (Winston and Horn, 1984), but Prolog (Clocksin and Mellish, 1987) is gaining in popularity. Lisp offers a lot of flexibility for writing rules. The expert system builder can specify his or her own framework of rules. Prolog is based on Horn clauses of logic. Rules need to be written explicitly in formal logic. Prolog has a built-in backward chaining mechanism, which makes it more convenient for some applications. However, caution needs to be exercised while using the back tracking mechanism in Prolog. Neither Prolog nor Lisp is very convenient for handling mathematical computations. Several expert system building tools are currently available. Examples are EMYCIN (Van Melle *et al.*, 1981), KAS (Duda *et al.*, 1979), EXPERT (Weiss and Kulikowski, 1979, 1981), OPS-5 (Forgy, 1981), ROSIE (Fain *et al.*, 1981), RLL (Greiner and Lenat 1980), Hearsay-III (Blazer *et al.*, 1980). Object-oriented programming has become popular in recent years. Smalltalk-80 and Objective-C are two important developments. NEXPERT is a shell used for building expert systems based on object-oriented principles.

After deciding on the programming language, the represented knowledge should be implemented in the language chosen. We describe the implementation aspects with reference to the Prolog environment. The three basic steps are:

- transforming the declarative and procedural knowledge to clauses in first-order predicate logic

- transforming the clauses from the first step to Horn clauses
- transforming the Horn clauses into Prolog expressions.

Consider the implementation of the declarative knowledge 'surface_finish of work piece ['p1' is 130 mu]'. This is represented in Prolog as a fact: surface_finish (p1, 130). Procedural knowledge or rules are implemented as follows. Consider the following rule:

Rule 1: **IF** the surface finish of a work piece > 120 mu
**THEN** it is rough
In Prolog this is represented as
rough (X):-surface_finish (X,Y),Y > 120.
In general Prolog rules are represented in the form
$P_0$:-$P_1$,$P_2$, . . . $P_n$
and read as
**IF** $P_1$, $P_2$, . . . $P_n$ are true
**THEN** $P_0$ is true.

Similar rules should be lumped together and, when applicable, rules should be linked together to improve processing efficiency.

One important consideration in the implementation phase is that of the user interface. Since the expert system is expected to help the novice user, it should be user-friendly. In addition, it should be conversational in its approach. When possible, visual graphics should be used. Explanations should be concise and clear and should follow the reasoning chain.

## Learning

It is very beneficial for an expert system to be able to use its experience in solving new problems. One of the important aspects of human intelligence is the ability to learn. This facet of human intelligence is not well understood. People know very little when they are born and absorb many things as they grow. The increase in knowledge is acquired by several methods. Organised knowledge is essential when incorporating learning into AI systems. Cohen and Feigenbaum (1982) classify four perspectives of learning as follows:

1. Learning as a process by which a system can improve performance.
2. Learning as a process through which explicit knowledge is acquired.
3. Learning as a process of acquiring skills.
4. Learning as a process of theory formulation.

The basic types of learning are dictionary learning, advice-taking, induction, learning by matching, learning by analogy and rule-based learning.

Dictionary learning is the simplest form of learning. Knowledge from the environment is supplied in a form directly usuable by the performance element.

Any computer program can have this type of learning. In a typical computer program the data are taken and stored and appropriate instructions are applied. The given facts are compared with the declarative base and appropriate actions are taken.

In advice-taking, knowledge is supplied from the environment to the system in a non-direct usable form. The AI system has to understand, interpret the higher-level advice and use it appropriately. For example, in a machining situation, advice could be 'turning process needs coolant'. The AI system designated to use this advice should be able to understand 'turning process' and 'coolant'. It should also know that the coolant is only needed when the machine is in operation. In this type of learning, advice is generally taken, 'operationalised' and put to use.

Learning by induction, in its simplest form, is learning by examples. Most of the AI research in learning focuses its attention on induction. Consider the procedural rule of the form **IF** (situation) **THEN** (action). Using this rule, given a situation, it is possible to generate an action. This action is already defined in the procedural knowledge. Learning by induction can be defined as the ability to generalise actions, based on situations. Charniak and McDermott (1985) describe the process of inductive learning as follows: 'The problem is to find a pattern or concept definition that applies to one situation and does not apply to another.'

Learning by matching is similar to dictionary learning. Facts about a general situation can be applied to a specific situation. General patterns are matched with a specific pattern and the corresponding variables in the general pattern are instantiated to the facts in the declarative knowledge.

In learning by analogy, similarities between different things are identified and appropriate knowledge is extracted by the system. It is possible for a new situation to be similar to an old situation, but not exactly the same. In such circumstances, the actions that are applicable in the old situation may be appropriate to the new situation. Very little work has been done in this area. The basic problems faced are:

- what is analogy?
- how can analogies be recognised?
- how can the analogous knowledge be used?

Rule-based learning is used in expert systems, since they need constantly to update their procedural rules. New rules need to be added and new knowledge must be acquired. The expert system can be programmed to learn new rules. However, this task is far too complicated. This is useful only when the rules are simple and straightforward, without nested levels. The major problem in rule-based learning is that of consistency enforcement. The new rules added, or the new knowledge acquired, should be consistent with the old rules and knowledge, and the overall configuration of the system. For example, if there is a rule to classify a department as belonging to the category of heat-producing departments, by analogy the system should be able to learn about classifying a department to the noise-producing class. This type of learning can be implemented with the help of rules.

After looking into the details of expert systems, we focus our attention on some manufacturing applications.

## *Expert systems in manufacturing*

In this section, by way of examples of applications in management, we will consider *some* (of the many) applications of expert system concepts in manufacturing. The systems illustrated will also show the reader how the concepts developed in the previous sections have been applied in different domains.

### Process planning

Process planning in the manufacturing environment requires a considerable amount of human expertise, and has been one of the areas of expert system applications. Several knowledge-based systems have been developed for process planning.

GARI (Descotte and Latombe, 1981) is a knowledge-based system for process planning in the 'metal-cutting' industry. Knowledge in GARI is represented by production rules, where the left-hand side consists of conditions about the part to be manufactured and the right-hand side contains pieces of advice representing technological and economical preferences. Each piece of advice is provided with a weight representing its importance. Since several pieces of advice can be conflicting, GARI is provided with the ability to resolve conflicts. Conflicts are resolved by selecting the piece of advice with the highest weight. Multiple plans can be generated by backtracking to the point of conflict, reintroducing a piece of advice that has been rejected, and proceeding with the planning. The knowledge base consists of a description of features, dimensional and geometrical relations between features, and global pieces of information, forming the declarative knowledge of the system. The manufacturing rules are of the **IF . . . THEN** type, and are parameterised by simple variables and set variables. GARI is implemented in the Maclisp language and operates on a CII-Honeywell Bull HB-68 computer under the Multics system. It is an experimental system and consists of a knowledge base of about 50 rules. Since the planning is dependent on weights, different weights can give significantly different plans.

TOLTEC, a process planning system that learns from experience, has been developed at Purdue by Kashyap and Tsatsoulis. It uses the concept of dynamic memory structures developed for understanding natural language, and various memory organisation structures on different levels of conceptual detail to store knowledge. These structures are (from highest to lowest level of abstraction):

- meta-MOP (memory organisation packages)

- MOP
- scenes
- scripts

Planning in TOLTEC proceeds from a skeletal solution to a more detailed one, by hierarchical decomposition of the solution. Constraints and memories of previous failures are used to guide the planner in the selection of appropriate solution instantiations in dynamic memory. TOLTEC uses its TOPs (thematic organisation packages) to learn from its past mistakes. TOLTEC can be guided by the user in selecting alternative solutions, or can automatically improve its performance by avoiding known manufacturing errors or by detecting inconsistencies between its solution and its deep knowledge (constraint conflicts). In all these cases TOLTEC reorganises its dynamic memory to avoid repetition of the same errors, or best to match the user's preference solutions.

## Scheduling

Expert systems can be of great help in scheduling activities in a 'jobbing' shop. Scheduling is a combinatorial problem and belongs to the class of NP-hard problems. An expert system can be developed that has the same knowledge and understanding of qualitative measures as the human scheduler.

ISIS-II (Fox, 1983*a, b*) is a constraint-directed reasoning system for scheduling factory job shops, and takes a heuristic approach to generating schedules. Knowledge consists mainly of constraints and the expertise of the scheduler. The expert knowledge can be used to relax the constraints and boundary solutions. This reduces the number of available alternatives and hence search becomes efficient to select the best alternative.

Constraints in ISIS are categorised into four types. First there are organisational constraints, which include job tardiness, work in progress, resource levels, cost production levels and shop stability. The second category is the physical constraints, and specifies what an object can or cannot be used for. The third category is the gating constraints, which define conditions to be satisfied before an object can be used or a process started, and consists of operation precedence and resource requirements. Preference constraints form the fourth category, and provide a means for expressing constraints. The constraints are represented as frames containing three slots: precondition, evaluation function and weight. The constraints are used to guide the search, and conflicting constraints can be relaxed. Relaxation is done in a generative fashion during heuristic search, or a set of rules is used to perform a post-search analysis. Scheduling decisions are made on the basis of current and future costs and profits. The ISIS system uses SRL for representation of frames and is programmed in Lisp.

Ben-Arieh *et al.* (1985) describe an experimental investigation into the routing of jobs in an automated production and assembly facility using a knowledge-based system. The objective of the system is to utilise all the data available in a computerised manufacturing cell, to create a good control mechanism to supervise the system and to generate real-time answers to problems that arise during system run time. The system consists of two types of knowledge: production knowledge as rules and procedural knowledge. A database is used to store the system states and consists of static and dynamic components. The static database contains data about the system queues, the part's current process, the time at which the part is required and the assembly states. There are three types of knowledge. The first type is the system behaviour knowledge stored in the database. The second type of knowledge is the analytical algorithmic knowledge that determines the routing decision according to current system states. The third type of knowledge is obtained by simulation. Conflict among selected rules is resolved by ordering the rules. The system is written in Prolog on a UNIX system and was tested in a simulated environment.

## Layout planning

Facilities layout is another candidate area for expert systems. They can be used to combine judgemental rules of human experts with quantitative tools in order to develop a good facilities design for a variety of unstructured situations.

IFLAPS (Tirupatikumara *et al.*, 1985) is a multi-criteria facilities layout analysis and planning rule-based system. As opposed to other systems, this system considers criteria other than materials flow—safety, employee convenience and constraint satisfaction. Knowledge is represented in logic predicates and implemented in Prolog. The system uses a combination of both forward and backward reasoning. The system uses augmented transition networks (ATN), originally applied to natural language processing, for generating alternative layout configurations. The conflicts among the objectives are resolved through rule-based heuristics and user interaction. The system explains its reasoning through the user interface and acquires new knowledge through rule-based learning.

One of the important inputs to the layout design is the flow data, which are generally dynamic in nature and are usually sensitive to time, product mix and qualitative measures. In general, a single global flow set cannot capture these dynamic changes. With multiple flow networks it is possible to compare different layout configurations. This can be further expanded by a functional design where functions are available that are analogous to flow sets. This allows the combination of flow networks into larger-scale networks.

## Maintenance and fault diagnosis

DELTA (Bonissone, 1983) is an expert system developed by the (American) General Electric Company for locomotive trouble-shooting. On selection of a particular fault area, the system proceeds with a series of detailed questions. At each step it explains the expert's reasoning. Finally, when the trouble-shooting system identifies the cause of malfunction, it generates specific repair instructions. DELTA consists of more than 500 rules, which are represented in a special representation language. The extendable language currently consists of nine predicate functions, eight verbs and four utility functions. It uses a hybrid forward/backward chaining mechanism for reasoning. DELTA was originally implemented in Lisp but later converted to Forth. It has been implemented on a microcomputer and has been field-tested.

Another area of expert system in the maintenance domain is the fault tolerance isolation and diagnosis in automatic test equipment (ATE). The rules used intuitively by humans must be articulated in a manner amenable to incorporation into ATE software. State measurements made by the ATE on the unit under test could then be used for defining a set of feasible hypotheses. Other knowledge, such as previous failure history, is relevant for determining a search strategy. Forest (Finin *et al.*, 1984) is one such expert system. It attempts to emulate experienced field engineers whose speciality is diagnosing faults undetectable by existing ATE software. Knowledge is represented as rules, and uses a frame-based representation. The rules are provided with weights to help in their selection. It also provides an explanation mechanism. The system is programmed in Prolog on a VAX 11/1971980 computer.

Current expert systems for machine fault diagnosis have a number of commonly recognised failings, including an inability to handle new faults, an inability to recognise when a fault is beyond the consultation system's scope, inadequate explanation of the final diagnosis, excessive requests for new information and difficulties in construction. The Consolidate system (Bublin and Kashyap, 1988) has been developed as an advance in the development of intelligent diagnostic consultation systems for electromechanical systems, and addresses these failings. It achieves this goal by merging a problem-solving method called heuristic classification with a representation of causal knowledge about both the normal and abnormal functioning of the system under investigation. Consolidate is implemented in Smalltalk-1980.

The system encodes a variety of knowledge types, including: heuristic knowledge (gathered through experience), which associates symptoms with faults; model-specific causal knowledge, which includes a representation of a system's structure, function and behaviour; generic causal knowledge, which describes constraints and requirements common to all systems; and control knowledge, which helps the system decide what task to pursue next. The system has been constructed to diagnose automotive carburettors, as this machine is representative of many of the features in mechanical domains.

## Expert database systems

We have seen a number of expert systems. However, attention must also be given to the development of a common database philosophy for factory automation. Some work has begun in this area, and should lead to the merging of existing relational database management systems and expert system applications (Kumara and Ham, 1988) and the possible development of intelligent manufacturing systems.

## References and Further Reading

Barr, A. and Davidson, J. (1980) Representation of knowledge. In Barr, A. and Feigenbaum, E. (eds.), *AI*. Computer Science Department report no. STAN-CS-80-793, Stanford University.

Barr, A. and Feigenbaum, E. A. (1981) *The Handbook of Artificial Intelligence Vol. 1*. Los Altos, CA: William Kaufmann.

Barr, A. and Feigenbaum, E. a. (1982) *The Handbook of Artificial Intelligence Vol. II*. Los Altos, CA: William Kaufmann.

Ben-Arieh, D., Moodie, C. L. and Nof, S.Y. (1985) Knowledge based control system for automated production and assembly. 8th International Conference on Production Research, Stuttgart, Federal Republic of Germany.

Blazer, R., Erman, L. D., London, P. and Williams, C. (1980) HEARSAY III: a domain independent framework for expert systems. Proceedings of the First Annual National Conference on Artificial Intelligence, Stanford University.

Bobrow, D. G. and Winograd, T. (1977) An overview of KRL: knowledge based representation language. *Cognitive Science*, **1**, 3–46.

Bonissone, P. P. (1983) DELTA: an expert system to troubleshoot diesel electrical locomotives. Proceedings of ACM, New York.

Brachman, R. (1979) On the epistemological status of semantic networks. In Findler, N.Y. (ed.), *Associative Networks: Representation and Use of Knowledge by Computer*. New York: Academic Press.

Bublin, S. C. and Kashyap, R. L. (1988) Consolidate: merging heuristic classification with causal reasoning in machine fault diagnosis. 4th IEEE Conference on Artificial Intelligence Application, San Diego, CA.

Buchanan, B. G. and Shortcliffe, E. H. (1984) *Rule-based Expert Systems*. Reading, MA: Addison-Wesley.

Charniak, E. and McDermott, D. (1985) *An Introduction to Artificial Intelligence*. Reading, MA: Addison-Wesley.

Chen, P. P.-S. (1976) The entity-relationship model—toward a unified view of data. *ACM TODS*, no. 1.

Clocksin, W. F. and Mellish, C. S. (1987) *Programming in Prolog*, 3rd Edition. Berlin: Springer-Verlag.

Cohen, P. R., and Feigenbaum, E. A. (1982) *The Handbook of Artificial Intelligence Vol. III*. Los Altos, CA: William Kaufmann.

Date, C. J. (1983) *An Introduction to Database Systems*. Reading, MA: Addison-Wesley.

Descotte, Y. and Latombe, J. C. (1981) GARI: a problem solver that plans how to machine mechanical parts. *Proceedings of the Seventh International Joint Conference on Artificial Intelligence*, Vancouver, Canada.

Duda, R. O., Gashing, J. and Hart, P. E. (1979) Model design in the prospector consultant system for mineral exploration. *Expert Systems in the Micro-Electronics Age*. Edinburgh: Edinburgh University Press.

Fain, J., Gorlin, D., Hayes-Roth, F., Rosenshein, S. J., Sowizral, H. and Waterman, D. (1981) *The ROSIE Language Reference Manual*. Technical Report N-1646-ARPA, Rand Corp., Santa Monica, CA.

Finin, T., McAdams, J. and Kliensosky, P. (1984) Forest: an expert system for automatic test equipment. Proceedings of IEEE Computer Science Conference on AI Applications.

Forgy, C. L. (1981) *The OPS5 User's Manual*. Technical Report CMU-CS-81-135, Computer Science Department, Carnegie Mellon University, Pittsburgh, PA.

Fox, M. S. (1983*a*) The Intelligent Management Systems: An Overview. Technical Report CMU-RI-TR-81-4, Intelligent Systems Laboratory, The Robotics Institute, Carnegie Mellon University, Pittsburgh, PA.

Fox, M. S. (1983*b*) Job Shop Scheduling: an Investigation into Constraint Directed Reasoning. Ph.D. Thesis, Carnegie Mellon University, Pittsburgh, PA.

Fox, M.S. (1985) Questionnaire on industrial expert systems. *SIGART News Letter*.

Goldstein, I. and Roberts, R. B. (1977) NUDGE: a knowledge based scheduling program. Proceedings of the Fifth International Joint Conference on Artificial Intelligence, Cambridge, MA.

Greiner, R. and Lenat, D. (1980) A representative language. Proceedings of the First Annual National Conference on Artificial Intelligence, Stanford, CA.

Hayes-Roth, F., Waterman, D. A. and Lenat, D. B. (1983) *Building Expert Systems*. Reading, MA: Addison-Wesley.

Kashyap, R. L., Smit, H. J., Tsatsoulis, C. and Wiggins, L. K. (1988) An intelligent system for integrating process planning and design. Proceedings of IEEE International Conference on Robotics and Automation.

Kumara, S. and Ham, I. (1988) Database considerations in manufacturing systems integration. *Robotics and Computer-Integrated Manufacturing*, **4,** 571–83.

Kumara, S. R. T., Joshi, S., Kashyap, R. L., Moodie, C. L. and Chang, T. C. (1988) Expert systems in industrial engineering. *International Journal of Production Research*. **24**, 1107–25.

Minsky, M. (1975) A framework for representing knowledge. In Winston, P. (ed.), *The Psychology of Computer Vision*. New York: McGraw-Hill.

Nau, D. S. and Chang, T. C. (1983) Prospects for process selection using artificial intelligence. *Computers in Industry*. **4,** 253–63.

Nilsson, N. (1971) *Problem Solving Methods in Artificial Intelligence*. New York: McGraw-Hill.

Quillian, R. (1968) Semantic memory. In Minsky, M. (ed.), *Semantic Information Processing*. Cambridge, MA: MIT Press.

Stefik, M., Aikins, J., Lazer, R., Benoit, J., Birnhaum, L., Hayes-Roth, F. and Sacerdoti, E. (1982) The organization of expert systems: a tutorial. *Artificial Intelligence*, **18,** 135–73.

Szolovits, P., Hawkinson, L. B. and Martin, W. A. (1977) An overview of OWL, a language for knowledge representation, Technical Report LCS-Tm-86, AI Laboratory, MIT, Cambridge, MA.

Tirupatikumara, S. R., Kashyap, R. L. and Moodie, C. (1985) Application of AI techniques to facilitate layout. Conference on Intelligent Systems and Machines, Department of Computer Science, Oakland University, CA.

Van Melle, W., Shortcliffe, E. H. and Buchanan, G. (1981) EMYCIN: a domain independent system that aids in constructing knowledge based consultation programs. Machine Intelligence, Infotech State of The Art report 9, no. 3.

Weiss, S. M. and Kulikowsi, C. A. (1979) EXPERT: a system for developing consultation models. Proceedings of the Sixth International Joint Conference on Artificial Intelligence, Tokyo, Japan.

Weiss, S. M. and Kulikowsi, C. A. (1981) Expert consultation systems: the EXPERT and CASNET projects. Machine Intelligence, Infotech State of The Art Report, 9. no. 3.

Winograd, T. (1975) Frame representation and the declarative procedural controversy. In Bobrow, D. and Collins, A. (eds.) Representation and Understanding, New York: Academic Press.

Winston, P. H. and Horn, B. K. P. (1984) *Lisp*, 2nd edition. Reading, MA: Addison-Wesley.

# 3.4 *Computer Simulation Methodologies and Applications*

*Frank W. Dewhurst*

## Introduction

In making decisions at either the strategic or operational and control levels, managers can call upon a variety of techniques—one of which, increasingly, is that of simulation (Beasley and Whitchurch, 1984). Simulation is an extremely powerful tool and has received a great deal of attention both from academics and practitioners because it has a wide applicability. Several studies (for example see Christy and Watson, 1983) have shown that there is a lack of awareness and knowledge of the technique, particularly among practising managers. This is perhaps partly due to the nature of the technique itself, in particular its diverse applicability and frequently confusing nomenclature. There is a need to increase acceptability of the technique (Christy and Watson, 1983). It should be made easier to use, less expensive and more accessible. This chapter considers these issues by outlining the nature of 'simulation', identifying the methodologies and presenting some of the typical applications. Recent developments that attempt to make the technique easier to use and a summary of some of the simulation software packages currently available are presented.

## The Nature of Computer Simulation

In all aspects of scientific study and analysis, model building has played an extensive role in the understanding of real-world phenomena. Using the simplest definition, a simulation model seeks to 'duplicate' the behaviour of a system under investigation and allows a study of the interactions among the components of a system by experimentation. The earliest simulation studies employed analogue models (i.e. small-scale physical models or maquettes), which can still be found in automobile, nautical and construction engineering (for example, in trying to determine the effects of weather on bridge design). Similar analogue models have been employed by designers of production systems when planning facility layouts. However, such analogue models are highly specific and require the physical alteration of the model, which can frequently be time-consuming and expensive. Furthermore, analogue models are often static, and cannot show how the various system factors interact dynamically. In many cases these analogue models have now been replaced by interactive digital computer graphics models, particularly for planning and design. Typical of such packages is Witness, details of which are given in Table 1. Such planning and design tasks are usually one-off, whereas management is frequently concerned with day-to-day (operational) or long-term (strategic) decision-making.

Assuming that managers faced with such decision problems do not wish to make 'seat of pants' judgements, various approaches are possible. In some cases it might be possible to conduct experiments on the real system. For example, traffic controllers might experiment with different traffic light phasings to achieve the most acceptable traffic flow at a particular road junction. In many decision-making situations such real-system experimentation is too risky to be undertaken and so a model of the system is constructed. In some cases it is possible to formulate a mathematical model of the decision problem and to employ a mathematical technique (e.g. calculus, linear programming, dynamic programming, etc.) to obtain a solution to the problem. Such mathematical models are often limited by the assumptions made in order that a specific technique can be applied to obtain the solution. For example, in analysing an inventory system, demand for items from stores is often assumed to be uniform, so that differential calculus can be employed to obtain the optimal ordering parameters. Even where such simplifying assumptions about a system are acceptable the solution procedures employed virtually restrict us to conclusions about the steady state of the system under investigation.

In many cases a steady state might not occur or may never be reached. For example, it may be necessary to study a system before it reaches a steady state. To illustrate this point, consider jobs arriving in a machine shop and the need to study how changes to the machines will affect queue formation into the shop. To apply the usual differential calculus arguments of queuing theory it is necessary to assume that steady state has been achieved so that the queue can be mathematically expressed in terms of the probabilities of arrival and service. However, a bottleneck may be causing an unacceptable level of queue formation and machine warm-up times may be a significant proportion of production run time, in which case the system will not have reached a steady state.

In some cases it may be possible to formulate a mathematical model of a system that cannot be solved except by some numerical procedure. For example, it may be possible to obtain a cost function for an inventory system in terms of the ordering parameters for each product, which cannot be solved explicitly. Typical in this respect are mathematical models that require the solution of differential equations or integrals. Experimentation with the parameters and evaluation of the inventory costs might allow the production manager to find acceptable values for the ordering parameters. In other cases, say in studying the effects of customer queues on a service system, it may not be

**Table 1** Proprietary products

| Name of package | Distributor | Computer system | Approx. price | Model employed and facilities |
|---|---|---|---|---|
| Aims | R. M. O'Keefe<br>Faculty of Mathematics<br>University of Southampton<br>Southampton<br>UK | Apple II | — | 3 phase event scheduling for discrete simulation |
| ECSL | Cle. Com<br>8 Stanley Road<br>Kings Heath<br>Birmingham B14 7NB<br>UK | DEC PDP11, DEC VAX | — | Activity scanning incorporates ISPG using an ACP |
| eLSE | Dr J. Crookes<br>University of Lancaster<br>Gillow House<br>Lancaster LA1 4YX | Apple II, IBM PC, IBM XT, AT, DEC VAX | — | Discrete 3 phase event scheduling |
| GASS | Acturial Micro Software<br>3915 A Valley Ct.<br>Winston Salem<br>NC 27016<br>USA | Apple II, IIe, III, IBM PC, XT IBM AT | $325 | Simulation of 10 variables in a user defined algorithm |
| GPSS/H | Wolverine Software Corp.<br>7630 Little River<br>Suite 208<br>Annandalf<br>VA 22003<br>USA | IBM 370, IBM AT, DEC VAX | — | Process interation, ISPG involved with editing and debugging facilities |
| GPSS/PC | Minuteman Software<br>P O Box 171<br>Stow<br>MA 01775<br>USA | IBM PC, IBM XT, IBM AT | $900 | PC version of above |
| Hocus | P–E Inbucon Ltd<br>Park House<br>Egham<br>Surrey TW20 0HW<br>UK | IBM AT | — | Continuous and discrete models |
| Interactive | Micro Simulation<br>50 Milk St<br>Suite 1500<br>Boston MA 02109<br>USA | Apple III, IBM PC, XT, IBM AT | $350–$750 | Process interaction for manufacturing and inventory models |
| InterSim | Decision Computing<br>3 St Edmund's Road<br>Canterbury CT1 2PD<br>UK | Apple II, IBM PC, XT, IBM AT | £435–315 | Discrete event scheduling with visual interaction |
| Microdynamo | Software Sales<br>Addison Wesley Pub. Inc.<br>Reading MA 01867<br>USA | Apple II, IBM PC, XT, IBM AT | $245–395 | Continuous system simulation |
| Micropassim | Dr C. Barnett<br>Dept of Physics<br>Walla Walla College<br>College Place<br>WA 99324<br>USA | Apple II, III, IBM PC, XT, AT, IBM AT | $125 | Discrete or continuous 3 phase event scheduling and process interaction |

| Name of package | Distributor | Computer system | Approx. price | Model employed and facilities |
|---|---|---|---|---|
| PC Model | Simulation Systems<br>SIMSOFT UK<br>6 Selsey Close<br>Hayling<br>Hants PO11 9SX<br>UK | IBM PC, XT,<br>IBM AT | £450 | Discrete event scheduling for manufacturing and assembly line models |
| Scheduling Simulator | L. Poizner<br>P O Box 2525<br>238 Davenport Road<br>Toronto<br>Ontario<br>Canada MSR 1JR | Apple, Lisa,<br>Macintosh | £535 | Discrete simulation for job shop scheduling |
| Witness (Seewhy) | Istel Inc.<br>Istel Ltd<br>Highfield House<br>Headless Cross Drive<br>Redditch<br>Worcs B97 SEQ<br>UK | IBM AT | $20 000 | Discrete graphical interactive simulation for production systems |
| SIMAN (Cinema) | Systems Modelling Corp.<br>Calder Square<br>P O Box 100074<br>PA 16805<br>USA | IBM PC, XT,<br>IBM AT | $200–1 500 | Discrete or continuous event scheduling, activity scanning process interaction special features for manufacturing |
| Simscript (II.5) | CACI<br>3344 NO., Torrey Pines Ct<br>La Jolla<br>CA 92037<br>USA | IBM PC, XT,<br>IBM AT | $250–24 500 | Discrete event scheduling and process interaction simulation |
| Slam II | Pritsker & Assoc.<br>P O Box 2413<br>W. Lafayette<br>IN 47906<br>USA | IBM PC, XT,<br>IBM AT | $200–975 | Continuous or discrete event scheduling, activity scanning and process interaction simulation |
| Solon | Y. J. Stephanedes<br>500 Pillburn Drive<br>Minneapolis<br>MN 55455<br>USA | Apple II,<br>IBM PC, XT,<br>IBM AT | $200–800 | Discrete graphical interactive simulation for transportation models |
| Tutsim | Micropacs<br>Graphics House<br>50 Gosport St.<br>Lymington<br>Hants SO4 9BE<br>UK | Apple II,<br>Apple III,<br>IBM PC, XT,<br>IBM AT,<br>DEC PDP11,<br>DEC VAX | £300–1 500 | Continuous graphical interactive for dynamic systems for process simulation |

All dollars are US dollars.

at all possible to model the decision problem mathematically. However, it might be possible to replicate the system by a logical model, typified by a flow diagram, employing operands such as **IF** (system is full) **THEN** (go to another or wait). Logical models of this type cannot actually be solved and would need to be enumerated from trial and error experimentation with the appropriate parameters.

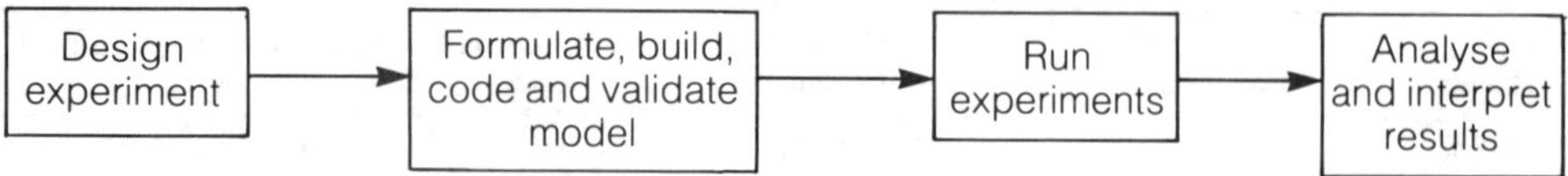


**Figure 1** A typical simulation exercise.

Either form of experimental approach to problem-solving is usually referred to as simulation and sometimes as the 'Monte Carlo method'. Both titles imply the use of a random sampling operation as part of the analysis but some authors have attempted to distinguish between them. Hammersley and Hanscomb (1964), for example, suggest that Monte Carlo should apply only to deterministic problems; that is, problems in which the data are known and fixed. The term simulation was reserved by them for decision problems of a stochastic nature. More recently the term Monte Carlo has been dropped in favour of simulation or simulation modelling. The former type of simulation, typically employed in the analysis of differential equations, is sometimes referred to as continuous simulation, whereas the latter form is usually referred to as discrete simulation.

Historically, the development of the digital computer cannot be separated from the development of the technique of simulation. Indeed the modern-day digital computer resulted from attempts to emulate digitally the behaviour of the electronic differential analyser and, in the 1960s, the term digital simulation was extensively used. Because of the profusion of these early digital–analogue simulators, a simulation software committee attempted to define the desirable features of a continuous system simulation language (CSSL). However, it is perhaps the more recent development of computers in discrete-event simulation that is of interest to most managers. Not only can the computer be used to perform the often tedious calculations required in a simulation experiment but also it can be used as a modelling device and, even more importantly perhaps, it can be used in the formulation of such models. Consequently we now find the terms computer-aided modelling and simulation (Spriet and Vansteenkiste, 1982) or computer-aided simulation and modelling (CASM) (Balmer and Paul, 1986) in use.

Simulation, then, can be seen simply as experimentation, usually via a computer, with a model of a system under investigation. Figure 1 typifies the overall approach.

## Simulation Applications

The major application of simulation has traditionally been as an aid to decision-making. Typical applications at the strategic level include facility planning, designing and planning of new plant and products, studying the effects on the organisation of policy changes, etc. Many simulation experiments have also been undertaken at the operational level (see, for example, Hurrion, 1978). Typical operational applications include production scheduling, maintenance scheduling, manpower planning, inventory analysis and queuing systems. Most books on simulation contain some typical applications. In some (for example Pidd, 1986) the cases described have been abstracted and the company names replaced by pseudonyms to protect their competitive position. In others (for example Poole and Szymankiewicz, 1977), actual case studies are presented. Numerous applications can also be found in the relevant journals and there is a specific journal of simulation, which frequently provides applications as well as reporting research in simulation.

An interesting and useful application for modelling procedural systems has been developed by Pritsker (1979), and is generically referred to as GERT. This type of problem-specific simulation will be of interest to many managers for scheduling and planning projects involving new technologies for which few historical data exist, and for which the usual program evaluation review technique (PERT) assumptions do not apply. Essentially GERT can simulate a project network and allow activities in the project to have times drawn from any distribution. In addition GERT allows for activities to be optional, for repetition of activities, for modification to activity durations and for any number of 'terminal' events in the network. Instead of seeking the critical (longest) path of activities, as in PERT, GERT identifies a criticality index for each activity. This is the relative frequency (percentage of time) with which an activity appears on the critical path after a large number of simulation runs of the project have been undertaken. There have been many applications of GERT to the planning of single and multiple projects, and Pritsker and Sigal (1983) provide a comprehensive list of such applications.

A second area of application is that of training, where simulation models are used for training operatives, managers and decision-makers in specific skills. This area of application is typified by the use of (analogue) flight simulators for the training of pilots and astronauts. Simulation, or simulation gaming as it is more frequently called in this context, is now being used successfully to train managers in a wide variety of organisations where real-time/on-the-job training is relatively expensive. There are journals and associations dedicated to this area of application. In the UK, SAGSET (The Society for the Advancement of Games and Simulation in Education and Training) produces the quarterly journal *Simulation/Games for Learning*[1]. In the USA, ISAGA (the International Simulation and Games Association) produces *Simulation and Games: An International Journal of Theory, Design and Research*, published by Sage[2].

Clearly any manager wishing to employ the technique of simulation will need to have knowledge of experimental design, model building, statistical analysis and computing. I will not discuss the issues of experimental design, statistical analysis or computing, which can be found elsewhere, but will consider the issues surrounding simulation model building.

## Simulation Methodologies

One of the advantages of simulation is that what may take weeks, months or years to happen in reality can be modelled in a few minutes of computer time. Therefore it is important to consider how time-flow is to be built into the model. One approach is to view the model in equal time intervals—such models are referred to as time-based. Obviously some decision needs to be made concerning the length of the time interval since if the time interval is too large changes in the system might be missed, while if the time interval is too small the computer will take longer to run through the simulation model and may not produce results in time for the decision to be taken. An alternative approach is to view the system when an event or change occurs rather than after a certain time has elapsed. This approach is usually referred to as event-based modelling, and is generally considered superior to the time-based approach.

In designing and building a model for a simulation experiment consideration needs to be given to the level of aggregation of the model. There is little point in producing an extremely detailed and accurate model if only crude estimates are required. Frequently a distinction is made in this respect between continuous and discrete simulation models. Generally, continuous models are used where the behaviour of the system under investigation depends more on the aggregate flow of events than upon the occurrence of individual events. Typically, continuous models are employed in highly complex systems where it is neither possible nor desirable to consider the fine detail or effects of individual events. However, the separation of discrete and continuous simulations is somewhat artificial and consequently a number of simulation packages allow the user to use discrete, continuous or mixed models.

Discrete-event simulation modelling is perhaps the most widely used form and it is therefore appropriate to consider it in more detail. There are four major and distinct approaches used in constructing discrete-event simulation models (DESMs) that have serious implications for the structure of the model

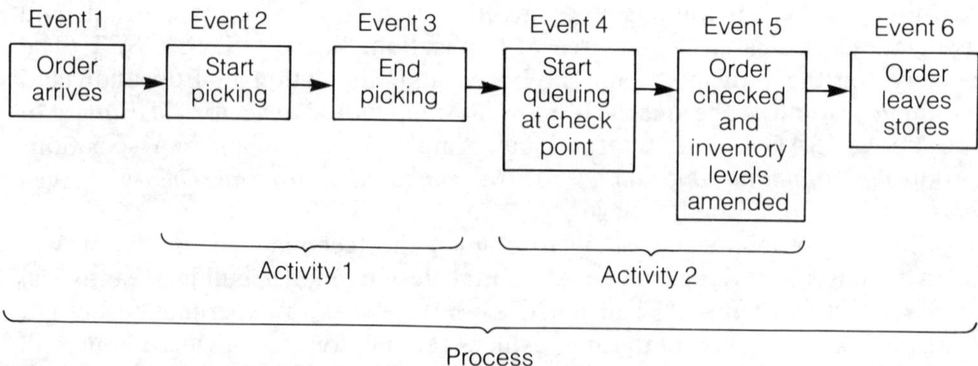


**Figure 2** A simple stores model.

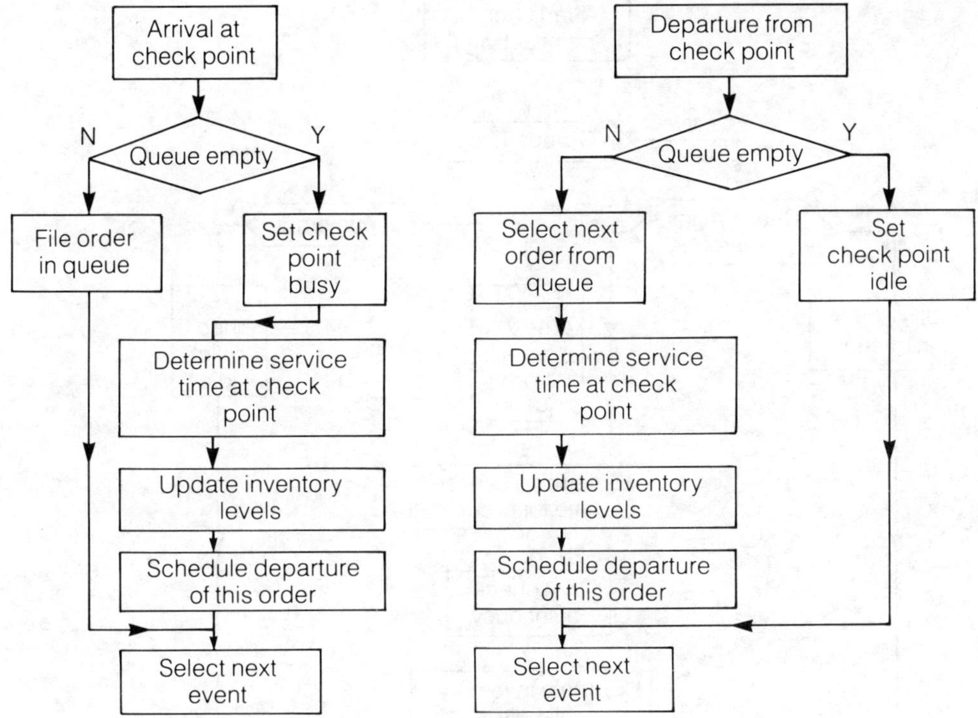


**Figure 3** Flowcharts for event scheduling at a check point.

and give rise to various discrete-event simulation languages, or more correctly interactive (simulation) program generators (ISPGs), from which the manager has to select the most appropriate for his or her purpose. An ISPG effectively interrogates the user about the problem to be modelled and produces code for a simulation program. These four approaches are usually referred to as:

1. Event scheduling, in which a complete description of any steps occurring when an event takes place are identified and coded as blocks.
2. Activity scanning, in which a review of all activities is undertaken to determine which can be started or terminated each time an event occurs.
3. Process interaction, in which the progress of an entity through a system is monitored from its arrival event through to its departure event.
4. The three-phase approach, which is essentially a combination of the event and activity approaches.

Figure 2 illustrates the first three approaches through a simple stores system. Each approach can be represented by a flowchart. The stores check point, events, activities and subprocesses are shown in flowchart form in Figures 3, 4, and 5 respectively.

In event scheduling, the event functions (e.g. start queuing at check point) are realised by event routines and events are scheduled by placing them in

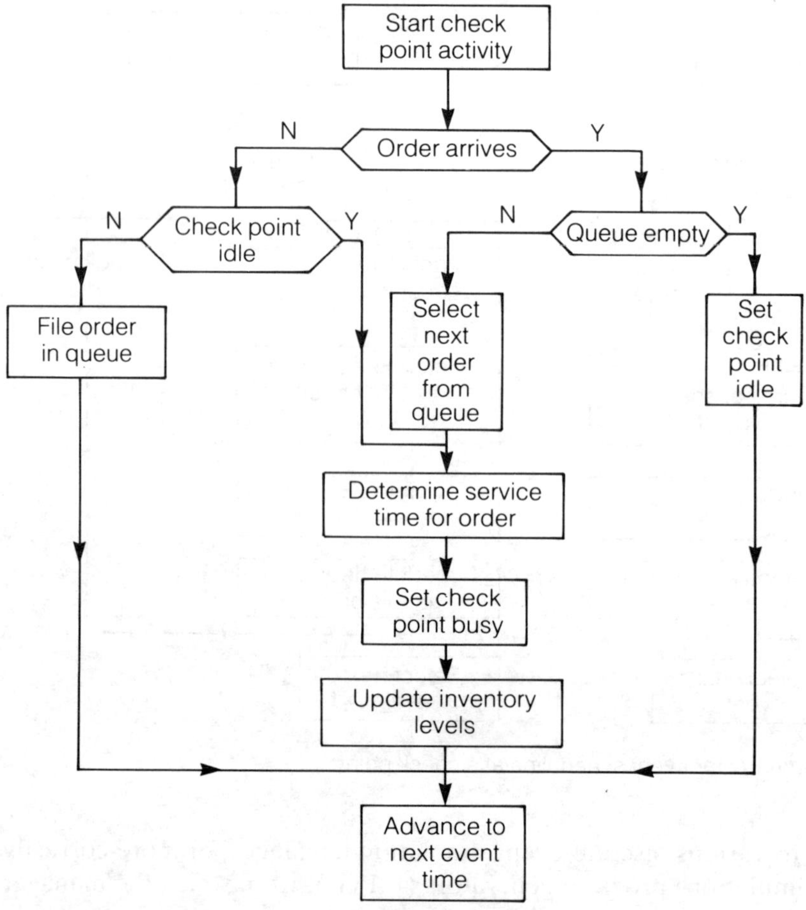


**Figure 4** Flowchart of activity scanning approach for a check point activity.

accordance with an event notice on the future event list. In contrast to the event scheduling approach, activity scanning requires that events be implicitly scheduled. Activities have conditions and actions, and at every time step the conditions are scanned in a fixed order. If the condition is found to be true the associated action is executed. This scanning, testing and execution continues until no further conditions are true and then the model is advanced to the next time step.

Process interaction can, to some extent, be viewed as a combination and refinement of the event scheduling and activity scanning approaches. A process is essentially a set of activities that are mutually exclusive, i.e. at most one activity can be activated at one time, and connected in the sense that the termination of one activity enables the initialisation of another. In contrast to the event scheduling and activity scanning approaches, time may elapse in the model. To overcome conflicts between overlapping processes, the process interaction approach uses wait, delay and advance statements in both

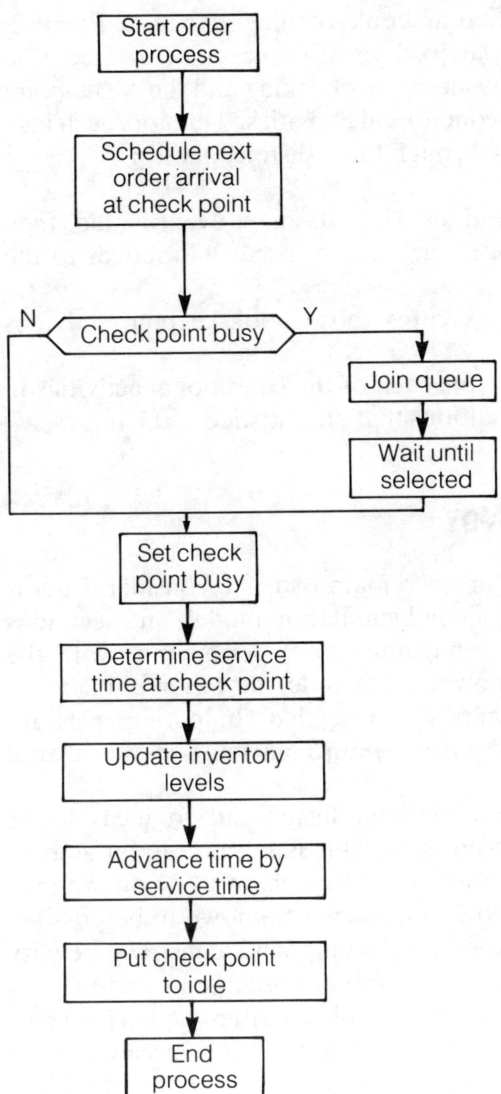


**Figure 5** Flowchart for process interaction approach at a check point.

conditional and unconditional contexts. The unconditional advance statement shares a common characteristic with the schedule statement used in the event scheduling approach. There is a difference, since each flowchart in the event scheduling approach represents only one event whereas several events can be included in a process interaction flowchart.

The three-phase approach combines the simplicity of activity scanning with the efficiency of event scheduling. Two types of activity are identified: book-keeping activities, which are started when their scheduled time is reached; and conditional activities, which depend upon some conditions being satisfied. In the stores example (Figure 2) events 1, 3, 4 and 6 depend only on time elapsing and so would be classified as book-keeping activities. Event 2, which depends upon an order having arrived and the picker being free, and event 5, which depends upon an order waiting for checking and the check point being free, would both be classified as conditional activities. This approach is so called because the simulation proceeds through three distinct phases:

1. The time-scan phase, which determines when the next event is due, then decides which book-keeping activities are due to occur and moves to the next event time.
2. The book-keeping phase, which executes those book-keeping activities identified as being now due.
3. The conditional phase, which attempts each of the conditional activities in the list and executes those with conditions that are satisfied.

## Selecting a Simulation Methodology

In choosing a suitable approach there are two main issues to consider. First is the time or cost incurred in producing a valid simulation model, and second is the time or cost incurred in testing, debugging and experimenting with the model. The former is clearly a manpower cost while the latter is mainly a computing cost. Although the trend appears to be that computing costs are declining and manpower costs are increasing, both contribute to the overall cost of applying simulation.

Generally, event scheduling simulations run faster and so incur lower computing costs than activity scanning models. This is because in an activity scanning approach all activities are scanned, even though only one or two may be possible, while in event scheduling only those events known to be possible will be executed. However, it is easier to develop and construct activity scanning models because they tend to produce modular segments and there is no need to be concerned about the sequence of activities at each event. Furthermore, it is easy to modify activity scanning models because extra activities can easily be inserted and existing activities can easily be modified. Thus the manpower costs will generally be less for activity scanning simulations although the computational costs will be higher than for event scheduling models. The process interaction approach has in its favour that it closely resembles the intuitive modelling that most managers and analysts apply to decision-making situations, but it is far more difficult to code or program than the event scheduling or activity scanning approaches and so requires much more sophisticated software. The process interaction approach therefore tends to be preferred by computer experts while the three-phase approach, which retains the simple programming of the activity scanning approach but has a running time more closely resembling the event scheduling approach, seems to be favoured in the UK (see, for example, Crookes, 1982; O'Keefe, 1984; Chew *et al.*, 1985; Pidd, 1986; Balmer and Paul, 1986). Other authors, such as Jennergen (1983), prefer the process interaction approach, but he also states that the choice of modelling approach in itself may not be as important as the selection of one approach, familiarisation with it and consistent use of it. Clearly, there is no best approach and a number of factors need to be taken into account when considering the use of simulation. A problem may lend itself to a particular simulation methodology or the choice of methodology might be constrained by the computer hardware and software available.

After selection of the most appropriate approach the model will usually be formulated in some structured manner through the use of flowcharts, activity cycle diagrams (ACP) or some special symbols, as in the general purpose simulation system (GPSS). When the logical model has been formulated and built it is necessary to code it to enable the experiments to be undertaken on a computer.

The earliest simulation models were coded in Fortran while more recently the programming languages Pascal and C have been used. Computer experts may prefer to use machine code or assembly languages but most managers would prefer neither low- nor high-level general programming languages—they may consider using a so-called simulation language (ISPG). However, if a new package has to be learned then the set-up costs increase and so it might be preferable to program the simulation in a language already known in the organisation. On the other hand, if a complex system that has a large pay-off is to be simulated, the project might warrant investment in a new package.

There are many simulation packages on the market offering a number of facilities, including ISPGs, debugging and editing facilities, random-number generators and a variety of random sampling distributions. Some packages provide immediate on-line interactive simulation experimentation while others operate in a batch mode. All provide some form of report with a statistical analysis, some in graphical form. Many of these are available for IBM PCs and compatibles which are within the budget of most managers. Table 1 lists some packages with names of suppliers, types of computer system required, an indication of the costs and the types of simulation model employed. Several of these simulation packages have been designed for production/manufacturing models while others have more general appeal. In choosing a suitable package the reader is advised not only to obtain further details from the supplier, to have demonstrations and obtain comments from existing users as with any software package, but also to consult the literature, where a number of these packages have been described in detail and evaluated on a comparative basis. Authors presenting such findings include Arthur *et al.* (1980), Crookes and Valentine (1982), Law and Kelton (1982), Mathewson (1985), Pidd (1986) and the entire issue of *Byte* of March 1984, which was devoted to simulation.

## Recent Developments

There are, at present, two major themes in the development of simulation: both result from developments in applications and computer technology.

The first of these developments is primarily concerned with improving the ease of use and applicability of simulation. Simulation languages and program generators (ISPGs), which facilitate the coding of simulation models, are now widely available. Many of these have been combined with statistical routines and report generators, details of which can be found in Table 1. Therefore the costs of implementing simulation should easily fall within the budgets of most managers.

The most time-consuming, and therefore costly, aspect of simulation is now that of model formulation. In the normal sequence of events it is necessary to define the problem and to build, test and validate the model before coding and running the simulation experiment. Managers will generally think of problems in day-to-day detail and in many cases might need to discuss the problem with an analyst or expert in order to develop the model. Some aids to model formulation are available as part of the simulation packages given in Table 1. However, it is still necessary for the manager to understand the nature of the model-building approach used and he or she would need to be familiar with and confident in identifying the events, activities and relationships of his or her particular problem to communicate effectively with an analyst or to develop the model. In order to use one of the existing packages he or she would need some experience in the use of flowcharts, activity cycle diagrams or some other form of symbolic representation before using one of the existing interactive model formulation aids. Often it is lack of experience in model building and formulation that results in invalid models, high development costs and, in some cases, loss of interest or even complete abandonment.

In an attempt to automate this formulation stage further a project entitled 'Computer Aided Simulation Modelling' (CASM) was formed at the London School of Economics. Under this project artificial intelligence techniques are being applied to aid model builders in the formulation stage before the use of ISPGs. Paul and Doukidis (1986) describe the use of a natural language understanding system that allows the model builder, through a discussion in English, to construct, validate, debug and edit the model. Haddock (1987) describes in ISPG that only requires knowledge about how the simulation and real system behave for the user.

Other work is presently being undertaken by a number of authors (for example Reddy and Fox, 1982) that will incorporate artificial intelligence techniques such as expert systems and intelligent knowledge-based systems into the framework of simulation. Calu *et al.* (1984) suggest that a simulation model must encapsulate knowledge about some aspects of the system under investigation while Shannon (1985) suggests that any simulation must contain knowledge about the problem domain in which the model is to be built, knowledge about the interpretation of results and statistics, and knowledge of the language used to code the simulation. Ford and Schroer (1987) have developed what they refer to as an EMSS (expert manufacturing simulation system), which effectively couples an expert system with a commercial simulation package (SIMAN) in order to study a medium-sized electronics manufacturing plant. This EMSS comprises a natural language interface that questions the user about the problem domain, a transformer that converts or interprets the natural language into an internal code, which is then passed to an understander that makes all the necessary inferences to complete the model before coding and a simulation writer that then translates this internal code into the Siman simulation code and executes the simulation. This EMSS also contains a rule-based expert system that makes recommendations to the user and the simulation writer to improve efficiency of the model. However, a full implementation of this EMSS has yet to be undertaken. The implications of this type of research are discussed by O'Keefe (1985, 1986). There are other possibilities in combining artificial intelligence with simulation, such as using expert systems to test model validity or using simulation to design expert systems, which have yet to be investigated.

The second of the recent developments is more concerned with developing new approaches to simulation modelling that result from advances in computer technology (hardware) rather than from the development in applications technology (software). Giloi *et al.* (1978) have developed an approach in contrast to the four approaches of event scheduling, activity scanning, process interaction and the three-phase method, termed parallel process simulation (PPS). This approach allows a notion of running time so that states are not observed at equidistant time instants or when an event occurs but at 'points of interest'. Central to this PPS approach is the computer programming language APL, which will presently restrict its application. However, such an approach would be extremely useful for investigating the typical resource contention problems faced by many managers.

An alternative, and not unrelated approach, is the concept of distributed simulation, in which a physical process is simulated by a network of tightly coupled computer processors. Roberts and Shires (1985) discuss this approach, in which individual computer processors emulate individual subprocesses of a complete manufacturing system. The communication lines between individual processors could represent flow patterns between the components of the system being modelled. The advantage of such an approach is in the high level of detail that can be incorporated into each computer processor module, so that experiments can take place at either the modular or system level. The disadvantages at present are reflected in the hardware limitations of networks and multiprocessor computers and, of course, in terms of the costs of the computing facilities required. However, as this type of computer technology advances it is expected that this approach will gain popularity. The approach has already been applied successfully to the investigation of a number of flexible manufacturing systems (FMS) reported by Charlish (1985) and at Normalair-Garrett, as reported in Wills *et al.* (1983) and Kellock (1985).

## Conclusions

Simulation is an extremely powerful tool for assisting managers in both strategic and operational decision-making and is becoming increasingly popular for training. The technique has developed in parallel with advancements in computer technology and will continue to do so. The availability of simulation packages for the microcomputer is now making the technique affordable to managers. The major problem faced by most managers is in relation to building a suitable model of the system to be investigated. Numerous texts on model formulation and simulation modelling are available (for example Pidd, 1986) and it is, perhaps, the lack of practical expertise in model building that is the major reason for the problems experienced in applying simulation. The *process* of modelling often provides valuable insights into a system and may in itself be sufficient in some cases to assist the manager in the decision-making process. There are a variety of modelling approaches that depend upon the nature of the problem under investigation. Present research is aimed at improving the modelling process, particularly for novices and inexperienced model builders.

## Notes

1. Details are available from the Centre for Extension Studies, University of Loughborough, Loughborough, Leics., UK.
2. Details can be obtained from Cathy Greenblat, Department of Sociology, Douglass College, Rutgers University, New Brunswick, NJ 08903.

## References


Arthur, J. L. *et al.* (1986) Microcomputer simulation systems. *Computers and Operations Research*, **13**, 167–83.

Beasley, J. E. and Whitchurch, G. (1984) OR education—a survey of young OR workers. *Journal of the Operational Research Society*, **35**, 281–8.

Balmer, D. W. and Paul, R. J. (1986) CASM—the right environment for simulation. *Journal of the Operational Research Society*, **37**, 443–52.

Calu, J. *et al.* (1984) Knowledge base aspects in advance modelling and simulation. Proceedings of the Summer Computer Simulation Conference, pp. 1247–53.

Charlish, G. (1985) Cincinnati showpiece on display. *Financial Times*, May 18.

Chew, S. T. *et al.* (1985) Three phase simulation modelling using an interactive program generator. CASM report, Dept of Statistics, LSE, UK.

Christy, D. P. and Watson, H. J. (1983) The application of simulation: a survey of industrial practice. *Interfaces*, **13**, 47–52.

Crookes, J. G. (1982) Simulation in 1981. *European Journal of Operations Research*, **9**, 1–7.

Crookes, J. G. and Valentine, B. (1982) Simulation in microcomputers. *Journal of the Operational Research Society*, **33**, 855–8.

Ford, R. D. and Schroer, B. J. (1987) An expert manufacturing simulation system. *Simulation*, **48**, 193–200.

Giloi, W. K. *et al.* (1978) APPL* DS: a powerful portable programming system for RT-level hardware description and simulation, microprogram and the simulation of parallel processing concepts. Technical Report, BF Informatik, Technical University of Berlin 78–21.

Haddock, J. (1982) An expert system framework based on a simulation generator. *Simulation*, **48**, 45–53.

Hammersley, J. M. and Hanscomb, D. C. (1964) *Monte Carlo Methods*. London: Methuen.

Hurrion, R. D. (1978) An investigation of visual interactive simulation with due dates and variable processing times. *Management Science*, **10**, 1264–75.

Jennergen, L. P. (1983) Simulation in microcomputers revisited. *Journal of the Operational Research Society*, **34**, 1053–6.

Kellock, B. (1985) Revival of the mini factories. *Machinery and Production Engineering*, **143**, 56–9.

Law, A. M. and Kelton, W. D. (1982) *Simulation Modelling and Analysis*. New York: McGraw-Hill.

Mathewson, S. C. (1985) Simulation program generators: code and animation on a PC. *Journal of the Operational Research Society*, **36**, 583–9.

O'Keefe, R. M. (1984) Programming languages, microcomputers and OR. *Journal of the Operational Research Society*, **35**, 617–27.

O'Keefe, R. M. (1985) Expert systems and operational research—mutual benefits. *Journal of the Operational Research Society*, **36**, 125–30.

O'Keefe, R. M. (1986) Simulation and expert systems. *Simulation*, **46**, 10–16.

Paul, R. J. and Doukidis, G. I. (1986) Further developments in the use of artificial intelligence techniques which formulate simulation problems. *Journal of the Operational Research Society*, **37**, 787–810.

Pidd, M. (1986) *Computer Simulation in Management Sciences*. Chichester: John Wiley and Sons.

Poole, T. G. and Szymankiewicz, J. Z. (1977) *Using Simulation to Solve Problems*. London: McGraw-Hill.

Pritsker, A. A. B. (1979) *Modeling and Analysis using Q-GERT Networks*. New York: John Wiley and Sons.

Pritsker, A. A. B. and Sigel, E. C. (1983) *Management Decision Making: a Network Simulation Approach*. Englewood Cliffs, NJ: Prentice-Hall.

Reddy, Y. V. and Fox, M. S. (1982) KBS: an artificial intelligence approach to flexible simulation. Technical Report CMU-R1-TR-82Y, Robotics Institute, Carnegie Mellon Unversity.

Roberts, E. A. and Shires, N. (1985) The application of multiprocessing to simulation. Proceedings of the First International Conference on Simulation in Manufacturing.

Shannon, R. E. (1985) Expert systems and simulation. *Simulation*, **44**, 275–84.

Spriet, J. A. and Vansteenkiste, G. C. (1982) *Computer-aided Modelling and Simulation*. New York: Academic Press.

Wills, K. F. *et al.* (1983) Advanced computer aided engineering and manufacturing. *Proceedings of the Institute of Mechanical Engineers*, **1978**, 81–9.

# Name Index

# Subject Index